TRANSACTIONS

OF THE

AMERICAN PHILOSOPHICAL SOCIETY

HELD AT PHILADELPHIA

FOR PROMOTING USEFUL KNOWLEDGE

NEW SERIES—VOLUME 45, PART 4
1955

A DOCUMENTARY HISTORY OF THE PROBLEM OF FALL FROM KEPLER TO NEWTON

De Motu Gravium Naturaliter Cadentium in Hypothesi Terrae Motae

ALEXANDRE KOYRÉ

Professor, Ecole Pratique des Hautes Etudes, Sorbonne

THE AMERICAN PHILOSOPHICAL SOCIETY

INDEPENDENCE SQUARE

PHILADELPHIA 6

October, 1955

Library of Congress Catalog
· Card No. 55–9770

A DOCUMENTARY HISTORY OF THE PROBLEM OF FALL FROM KEPLER TO NEWTON

De Motu Gravium Naturaliter Cadentium in Hypothesi Terrae Motae

ALEXANDRE KOYRÉ

CONTENTS

I. INTRODUCTION

It is well known that it was Hooke's invitation to Newton (of November 24, 1679) to resume his scientific correspondence with the Royal Society, or, more exactly, Newton's reply, attempting to determine the trajectory of a heavy body falling from the summit of a high tower, followed by Hooke's criticism, and the ensuing polemics, that turned Newton's mind away from "other business" and back to the study of terrestrial and celestial mechanics.[1]

It is less well known that the problem discussed by the two great scientists—the trajectory of a falling body *in hypothesi terrae motae*—had behind it a rather long, complicated and extremely interesting story.[2] A story that reveals to us some of the psychological and epistemological obstacles that lie in the path of the new science of the seventeenth century; that shows us how difficult it was even for such revolutionary minds as Galileo and Newton to free themselves from the conjoint influence of tradition and common sense, to draw—and to accept —the inevitable consequences of their own fundamental concepts. To accept, for instance, that in a certain sense —though in a sense quite different from his own— Aristotle was right in his classical objection against the motion of the Earth; namely, that—Copernicus and Galileo notwithstanding—things would (or do) happen

on a moving Earth that do (or would) not occur on a still-standing one; even more right than he knew, and that not only a stone dropped from the top of a tower will never fall at its foot, but even worse, that a freely-falling body would never—even if the Earth were permeable—reach its center, but would for ever turn around it.

Nobody, with the exception of Borelli, could even grasp such an absurd consequence; and nobody—at first not even Newton—could accept it as true. This, much more than the inner difficulty of the problem, prevented men like Galileo and Fermat, Mersenne and Borelli from reaching the correct solution of the problem of the determination of the trajectory of a falling body *in hypothesi terrae motae*.

The story of their attempts to solve it—and their errors—is thus the story of the relentless struggle of the human mind against itself. A sad, and an exciting, story.

It is this story that I will try to unfold in the following paper, and, as the texts I shall be dealing with belong, with very few exceptions, to long-forgotten authors and are buried in books available only in the very largest and oldest libraries,[2a] I shall quote extensively the representatives of this *ignota litteratura*. It is, in my opinion, not only very useful, but even indispensable, to put the "documents of the case" before the readers.

II. KEPLER, LOCHER, GALILEO

As a matter of fact, Newton, in his first letter to Hooke, did not attempt to deduce the complete trajectory of a body falling to the (moving) Earth from a fixed point above its surface. He only wanted to explain to him the theoretical background of the experiment he proposed to the Royal Society and his expectations concerning its results, namely, that this body would fall to the east (and not to the west) of the tower.

Nevertheless, he told Hooke that the curve described by the falling body would be a *spiral*. Moreover, he added to his letter a drawing of this spiral that clearly shows it terminating in the center of the Earth (fig. 1).

It has been advanced by J. H. Hagen[3] that "The idea of drawing this spiral may have been suggested [to

---

[1] *Cf.* W. W. R. Ball, *An Essay on Newton's Principia*, London, Macmillan, 1893; Jean Pelseneer, Une lettre inédite de Newton, *Isis* 12 (38), 1929; and my paper, An unpublished letter of Robert Hooke to Isaac Newton, *Isis*, 1952.

[2] On the history of the problem (after Newton), *cf.* A. Armitage, The deviation of falling bodies, *Annals of Science* 5, 1947.

[2a] Thus neither the British Museum nor the Bibliothèque Nationale possesses the complete collection of these works.

[3] J. H. Hagen, *La Rotation de la terre, ses preuves mécaniques anciennes et modernes*, II, Rome, 1911: "L'idée de dessiner cette spirale peut lui avoir été suggérée par un passage de l'*Epitome*

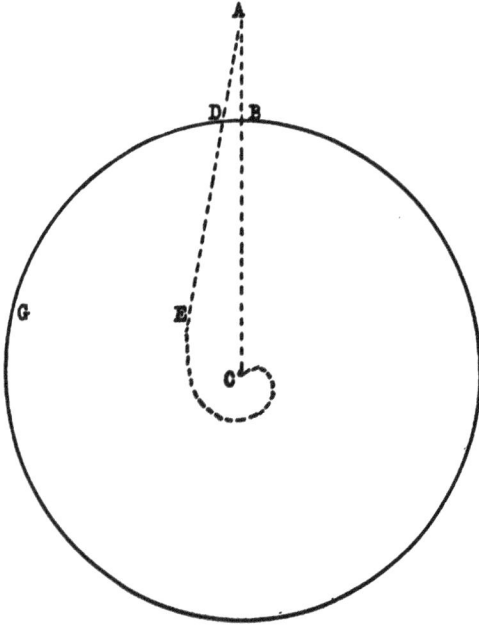

FIG. 1

Newton] by a passage of the *Epitome Astronomiae Copernicanae* of Kepler ... where a curve, resembling a spiral is represented." This of course is quite possible, though, as we shall see, Newton could have been inspired by more recent drawings. As a matter of fact, Kepler's curve is (*a*) not a spiral, (*b*) does not reach the center of the Earth, and (*c*) represents in a very rough manner the trajectory of a body falling, not from a point on the Earth, but from a point in the sky.[4] The

meaning of this drawing, which Hagen strangely enough considers as "not very clear," is thus quite different from that of Newton. It purports to show that a body falling to the Earth will not describe a straight line (as Aristotle thought it should) or an undulating one, as some one (perhaps Locher) asserted. This body, being subject to the influence of the *quasi*-magnetic (attractive) power of the Earth (*raptus*), will follow it, because it is drawn by the Earth and drawn more strongly as it comes nearer to it. This implies that when the body reaches the Earth it will fall neither on the spot which *was* directly underneath its point of departure at the moment it started its motion—this spot is now far away to the east—nor on the spot which is *now* directly underneath the starting point of the motion—this one will be to the west—but between the two. In other words, the falling body will "lag behind" the moving Earth, though not as much as the Aristotelians think. It will certainly not outrun its motion.[5]

Kepler's considerations and drawing, even if they did not have a direct influence on Newton, show us in any case that the problem of determining the real—curvilinear—path of a falling body was acute at the beginning of the seventeenth century.

M. Cornelis De Waard, in one of his very excellent introductions to the *inedita* of Fermat, points out another study of the same problem, contemporary and even somewhat earlier than that of Kepler.[6] It is a study which, following a tradition that starts with Galileo, he ascribes to Christopher Scheiner and to which, quite rightly, he attributes a great importance because (*a*) it puts forth the idea that the falling body will describe a spiral (and even different spirals according to its position on the equator or on some parallel), and (*b*) because it is quoted by Galileo, who should have learned something from it but did not.

As a matter of fact, the author of the historically very interesting *Mathematical Disquisitions* to which Cornelis De Waard has drawn our attention was not Christopher

---

Astronomiae Copernicanae de Kepler, paru en 1618 (*Opera omnia*, VI, 1, 1865, p. 182) où est représentée une courbe ressemblant à une spirale. Le sens que Kepler attachait à cette courbe est assez incertain, mais pourrait peut-etre s'éclairer par les passages d'auteurs antérieurs à Copernic. Léonard de Vinci s'était occupé du problème suivant: 'Du grave descendant dans l'air; les éléments étant animés d'un mouvement de circonvolution dont l'entière revolution a lieu en 24h.' Le texte est accompagné d'une sorte de spirale qui commence 'de la partie la plus élevée de la sphère du feu' et se termine à la surface de la terre. Le but de cette figure est de montrer que le trajectoire du grave, bien que rectiligne, absolument parlant, prend en apparence la forme d'une spirale, par suite de la rotation des sphères célestes. La spirale de Léonard est reproduite et expliquée dans un ouvrage récent de P. Duhem, *Etudes sur Léonard de Vinci; ceux qu'il a lu et ceux qui l'ont lu* (Paris, 1908), pp. 252–255." *Cf.* J. Pelseneer, Une lettre inédite de Newton, *Isis*, 12 (38): 241, n. 15, 1929.

[4] This is the text which Hagen has in mind: J. Kepler, *Epitome Astronomiae Copernicanae*, 182, *Opera Omnia*, ed. Frisch, Frankoforti, 1865, *Quae est ergo genuina figura motus gravium respectu spatii mundani?*:

"Quidam sedulus astronomiae cultor, sed non satis consideratus, pingit casum lapidis versus terram cis et ultra perpendi-

culum serpentinis flexibus fluctuantem, ut flexus numeri respondeant gyrationibus telluris, interim dum lapis in casu est; nec perpendit quod lapis desertus a partibus Terrae, quibus erat initio perpendicularis; veniat in raptum succedentium vicinarum partium, semper in illam plagam deflexo lapsu, in quam volvitur Tellus, initio parum, in fine magis magisque, quia raptus ex propinquo est fortior.

"Itaque figura motus gravium, si eorum aliquid ex coeli loco remotissimo versum Terram, in une certo loco rotatam, decideret, esset propemodum iste qualis his rudi Minerva depictus est, ubi circulo Terrae in 14 partes diviso, linea casus in totidem sed inaequales, supra breves, infra longiores, partes circuli ordine, trahendi munere defunctae, ad sua pristina loca redierunt, tres solum residuae, praeventae fine lapsus, non traxerunt perpendiculariter." *Cf.* fig. 2.

[5] On Kepler's theory of attraction and *raptus*, *cf.* my *Etudes Galiléennes* 3, *Galilée et la loi d'inertie*, 28 ff., Paris, Hermann et Cie, 1939.

[6] *Cf. Œuvres de Fermat*, supplément aux tomes I–IV, documents inédits publiés avec notices sur les nouveaux manuscrits par M. C. de Waard, "Ecrit anonyme sur la spirale de Galilée, son attribution à Fermat," 1, Paris, Gautiers-Villars, 1922.

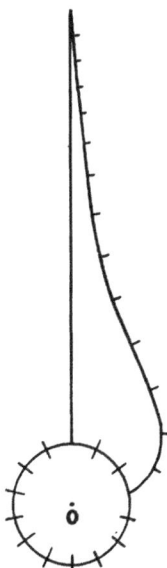

FIG. 2

Scheiner, who only presided over the *Disquisitions* (held in 1614), but his pupil Johann Georg Locher.[7] Besides, he did not positively affirm that a body falling to the Earth or, more exactly, to the center of the Earth, will actually describe a spiral (in a plane or on a cone). He only asserted this hypothetically, stating that such would be the path of the body *if the Earth moved*, and used this consideration as an objection against the Copernicans. Moreover, in determining this hypothetical path of the falling body, he made some rather heavy blunders, which made him the object of Galileo's biting sarcasm.[8] Since Galileo's time no one, probably not even Father Mersenne or Father Riccioli, though they both quote them, has ever so much as looked at the *Disquisitions*. Indeed not only do they—following Galileo—ascribe the *Disquisitions* to Scheiner,[9] but they even seem to borrow their quotations from the

[7] Here is the full title of the work of Johannes Locher: *Disquisitiones mathematicae de controversiis et novitates astronomicis. Quas, Sub Praesidio Christophori Scheineri, De Societate Jesu, Sacrae linguae et matheseos in Alma Ingolstadiensi Universitate, Professore Ordinarij, Publice Dusputandas, posuit propugnavit Mense Septembri, Die . . . Nobilis et Doctissimus iuvenis, Johannes Georgius Locher, Boius Monacensis, Artium et Philosophiæ Baccalaureus, Magisterii Candidatus, Iuris Studiosus*, Ingolstatii, 1614.

[8] *Cf.* Galileo Galilei, *Dialogo dei due massimi sistemi del mondo*, 145–149. *Opere*, Edizione Nazionale 7.

[9] It is possible, of course, that the work of Johann Locher, a pupil of Scheiner, was also inspired by him. It is even probable that such was the case, and this would explain why Galileo ascribed the *Disquisitions* to Scheiner himself, thus making fun of the master rather than of the pupil.

Dialogo. We have, therefore, to restitute the work to its author and to quote *verbatim* the passage in question.

Having explained that a falling body must necessarily follow a straight path toward the center of the Earth (of the universe) Locher continues: [10]

These [propositions] being posited, it is necessary that on a circularly moving Earth all [the things] that are placed perpendicularly above it in the air should be carried around with it by a motion proportional in all [respects]. Therefore when the Earth BNX [fig. 3] is turning round on the axis $\nu\lambda$, from B through X to B, the small sphere A which is placed directly above the point B, and the small sphere Υ above the point ι, and the small sphere ν above the pole would be likewise turning around, and would accomplish their circuit in the same time as the points of the earth B, ι, and the pole which are beneath them. And if we should admit that these balls, equal in weight, size, gravity and placed in the concave (to concede to our adversaries as much as possible) of the moon, would be allowed to descend freely, then if we assume that the motion downwards be equal in speed to the circular motion {which, as a matter of fact, is by no means so, as the ball A placed on the concave of the moon would in its hourly circular motion traverse at the minimum 12,600, twelve thousand six hundred, German miles, and the ball Υ in our climate 8,431, eight thousand four hundred and thirty-one miles, these figures being deduced from the teaching [a] of Copernicus and [b] even more from that of his follower Maestlin (to which motion it is however impossible that the descent be equal in speed)}, it would take at least six days: during which time the ball A would six times turn around the Earth and with the Earth, in order that it always be directly above its point; and as many times the ball Υ, which in twenty-four hours would arrive in α, would move from Υ to Z in twelve hours, and from there finally to α, wherefrom back to β and therefrom to γ in one day; and thus would turn in accordance with the Earth from γ to ϵ

[10] *Disquisitiones mathematicae*, 29: "His positis; necesse est terra circulariter mota omnia ex aëre eidem *perpendiculariter* imminentia, cum eadem circumferri motu per omnia proportionali. Igitur dum terra BNX in axe $\nu\lambda$ circumvolutatur, a B per X in B; interim sphaerula A puncto B, et sphaerula Υ puncto ι, et sphaerula ν polo terrae incontinenter super imminens, etiam circumagetur, eodemque tempore periodum absoluet quo puncto terrae B, ι et polus, illis subiecta. Quod si hasce pilas aequales ponemus pondere, magnitudine, gravitate, et in concavo sphaerae Lunaris positas, libero descensui permittamus, si motum deorsum aequemus celeritate motui circum (quod tamen secus est, cum pila A apud concavum Lunae motu horario circularis confectura sit milliaria germanica minimum 12600, duodecies mille sexcenta, in nostro vero climate Pila Υ milliaria 8431 octies mille, quadrigenta, triginta et unum, ex sententia Copernici, [a] et multo magis [b] Maestlini eius sectatoris; (cui tamen cursui impossibile est celeritate aequari descensum rectum, etc.) elabentur minimum (ut multum cedamus adversariis) dies sex: quo tempore sexies circa terram et cum terra in aëre libero vertetur pila A, eo quod semper insistat puncto suo B, et toties pila Υ: quae dum viginti quattuor horis perveniat in α circulata est ex Υ in Z, horis duodecim, et ab hoc demum in α quo rursus tendit in β, et ex hoc in γ uno die; et sic cum terra versabitur proportionaliter, ex γ in ϵ per δ ex ϵ in η per ς; ad θ, ad ι, ad κ, ad C in Centrum universi. Eodem fieret in simili proportione, cum avibus, fumo, igne, nubibus, aliisque in aere libro ad horas integras subinde immobilis pendulis."

Locher supports his reasoning with a drawing, which Galileo does not reproduce, and which is rather interesting. We are giving it here somewhat simplified.

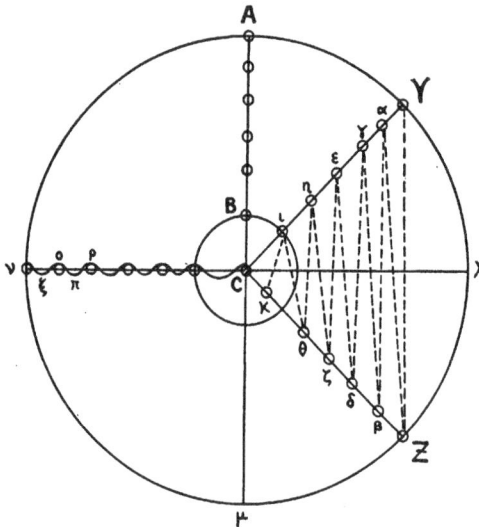

·Fig. 3

through δ, from ε to η through ζ, to θ, to ι, to κ towards C in the center of the world. And the same would occur in a similar proportion to birds, smoke, fire, clouds, and all other things that for entire hours remain motionless in the free air above the Earth.

But, thinks Locher, there is obviously no reason why bodies should behave in so strange and unbelievable a manner, and so he concludes with a rhetorical question: [11] "Why does the center of the sphere *A* falling down describe a spiral in one circle only, the center of the sphere Υ trace the same spiral line on a cone, the center of the small sphere ν trace a straight line congruent to the axis, meanwhile by its extreme point ν describing a screw-formed line contained on the surface of a cylinder?"

Locher's reasoning is a curious mingling of acuteness and misunderstanding.[12]   Still, in trying to deduce the consequences of the Copernican position, he certainly did show a rather unusual acumen, and Galileo, no doubt, should have paid more attention to him. Unfortunately, as we know, Galileo did not. He neglected even to discuss the Locherian conception, though

[11] *Disquisitiones mathematicae*, p. 33: "Quare centrum spherae A delapsae *spiram describit in unico circulo? Sphaerae* Υ *centrum eandem spiralem lineam designat in cono? Spherulae* ν *centrum, rectam lineam axi congruam designat, ipsa extremo puncto suo* ν, lineam giralem decurrens, in superficie cylindrica consignatam?"   (The italics are mine.)

[12] Such as the inane pleasure taken in calculating the (hypothetical) speed of a ball placed in the concave of the moon, a speed equal to that of the moon's concave in the hypothesis *of a non-moving Earth*; or the assumption of a fall which is vertical relatively to the surface of the moving Earth; or the miscalculation of the duration of the fall, etc., etc.

he put a very careful and perfectly exact account of it in the mouth of Simplicio. Led obviously by the rather numerous blunders and even quite gross errors (such as the idea that a body falling to the Earth from the concave of the moon, even if it fell with the same speed with which the sphere of the moon turns around the Earth in a diurnal motion, would need at least six days in order to reach the surface of the Earth [13]) of the *Disquisitions* into making them his *tête de turc* in his criticism of the anti-Copernican arguments against the motion of the Earth, he ridicules the work as a whole. Besides, could anything good come from the pen—or from a pupil—of Christopher Scheiner? Consequently, during the discussions of the second day of the *Dialogue on the two greatest systems of the world,* Simplicio, after having mentioned the existence of a marvelous booklet that completely overthrows the lucubrations of the Copernicans, proceeds to report its arguments, some of which, he says, are quite new. Thus, for instance: [14]

I will begin therefore with the objections which I find in the Treatise of Conclusions, . . . In the first place then, he bestoweth much paines in calculating exactly how many miles an hour a point of the terrestrial Globe situate under the Equinoctial, goeth, and how many miles are past by other points situate in other parallels: and not being content with finding out such motions in horary times, he findeth them also in a minute of an hour; and not contenting himself with a minute, he findeth them also in a second minute; yea, more, he goeth on to show plainly, how many miles a Cannon bullet would go in the same time, being placed in the concave of the Lunas Orb, supposing it also as big as *Copernicus* himself representeth it, to take away all subterfuges from his adversary.

But having made this most ingenious and exquisite supputation, he showeth, that a grave body falling from thence above would consume more than six dayes in attaining to the centre of the Earth, to which all grave bodies naturally move. Now if by the absolute Divine Power, or by some Angel, a very great Cannon bullet were carried up thither, and placed in our Zenith or vertical point, and from thence let go at liberty, it is in his, and also in my opinion, a most incredible thing that it, in descending downwards, should all the way maintain itself in our vertical line, continuing to turn round with the Earth, about its centre, for so many dayes, describing under the Equinoctial a Spiral Line in the plain of the great circle itself: and under other Parallels, Spiral Lines abouth Cones, and under the Poles falling by a simple right line. He, in the next place, establisheth and confirmeth this great improbability by proving, in the way of interrogations, many difficulties impossible to be removed by the followers of *Copernicus;* . . .

[13] It would, in this case, need less than four hours, as Galileo very nastily points out; *cf., Dialogo, loc. cit.*

[14] *Dialogo,* Ed. Naz., 7: 145. I quote from the English translation by T. Salusbury, *Dialogue on the Two Greatest Systems of the World,* 195, London, 1662. *Cf.* now, Galileo Galilei, *Dialogue on the great world systems,* in the Salusbury translation, revised, annotated and with an introduction by Giorgio de Santillana, 234 sq., Chicago Univ. Press, 1944. I am quoting the Salusbury translation as it was this one that was used by Hooke, Newton, and their contemporaries.

Simplicio's account is perfectly correct, as one can see from Locher's text which I have quoted.[15] As an argument against the Copernican doctrine it is obviously worthless because it is based on two false assumptions: (a) that, if the Earth turned, the whole sublunar world would turn with it, and (b) that in this case a body falling to the Earth, in spite of its describing in absolute space a plane or a conical spiral, would nevertheless in relation to the Earth fall in a perfectly straight line (the vertical). Yet there is no reason to dismiss it as lightly as does Galileo who quoting—or rather somewhat misquoting—Locher's final rhetorical question:[16] "Quare centrum spherae delapsae sub aequatore spiram describit in ejus plano? sub aliis parallelis spiram describit in cono? sub polo descendit in axe, lineam gyralem decurrens in superficie cylindrica consignatam?" replies:

Because of the lines drawn from the Centre to the circumference of the sphere, which are those by which *graves* descend, that which terminates in the Aequinoctial designeth a circle, and those that terminate in other parallels describe conical superficies; now the axis describeth nothing at all, but continueth in its own being. And if I may give you my judgment freely, I will say, that I cannot draw from all these Queries, any sense that interfereth with the motion of the Earth; for if I demand of this Author (granting him that the Earth doeth not move) what would follow in all these particulars, supposing that it do move, as Copernicus will have it; I am very confident, that he would say that all these effects would happen[17] that he hath objected as inconveniences to disprove its mobility; so that in this mans opinion necessary consequences are accounted absurdities: but I beseech you, if there be any more, dispatch them, and free us speedily from this wearisome task.

Locher's hypothetical solution of the problem of the trajectory of heavy bodies falling to the center of the Earth (if the Earth did move) was of course rather bad. But one may doubt whether the solution is much better which Galileo himself presented, at first tentatively in his letter to, or rather against, Francesco Ingoli, and later in his *Dialogue* where he stated positively that it was, if not true, at least very near the truth.

To refute the objections raised by the anti-Copernicans[18] was indeed perfectly easy. Copernicus himself, in order to explain why—if the earth moves—bodies falling from the top of a tower (or thrown up) apparently follow a straight line and do not "lag behind," had to make an appeal to the "community of

nature" of all the bodies constituting the earth and to the natural character for all of them of the circular motion which he attributed to the Earth. Kepler had to explain the apparently rectilinear trajectory of fall by the action of the attractive power of the Earth, by which all bodies are drawn in a violent *raptus* to which they unwillingly submit. However, according to the new mechanics developed or, if one prefers, inaugurated by Galileo, that of the principle of inertia, a body departing from a moving starting point conserves the motion it shared before departing. Thus, falling from the tower, it would—if the Earth moved—move not only *downwards* but also *to the east,* just as the tower itself. Consequently, it would come down exactly on the same spot to which it would fall if the Earth did not move. In other words, the motion of the falling body, relative to the tower and to the Earth, would be identical.[19]

But what about its absolute motion, or in other words the true path of the body? Indeed, on a horizontally moving platform, assuming moreover that the direction of gravity would be the same on all points of the platform (an Archimedean assumption which, though theoretically false,[20] is correct for all practical purposes), the trajectory, as Galileo already knew at the time of the publication of the *Dialogue*,[21] will be a parabola. But what if, as it is in reality, the motion of the tower is not horizontal but circular, and if the directions of gravity are not parallel but concurrent in the center of the Earth? According to Galileo, the motion in this case will be a semicircular one. Instead of going straight down to the center of the Earth, as it would, if the Earth stood still, a heavy body on a moving (rotating) Earth would combine its movement down (towards the center of the Earth) with its movement east and thus describe a semicircle with the Earth's radius as a diameter.

In so doing it would reach the center of the Earth in exactly the same time as it would if the Earth did not move. Moreover—and Galileo seems to be extremely glad and even proud to have discovered it—the length of the trajectory it would describe in its fall would be exactly the same as if it did not move at all, but remained at rest on the top of the tower. But let Galileo speak for himself:[22]

---

[15] Locher, however, does not mention the *cannon* ball, nor does he speak about an angel transporting it to the concave of the moon.

[16] *Dialogo,* 214, 237; p. 218 of the Salusbury translation; p. 259 of the Santillana edition. *Cf.* note 11, where the actual text of the *Disquisitiones* is quoted.

[17] This assertion of Galileo is rather strange because, of course, the "effects" deduced by Locher—for instance the body's remaining on the vertical—would *not* occur; the body, as Galileo quite correctly points out, would outrun the Earth; *cf. Dialogo,* p. 259; Santillana's ed. p. 249.

[18] *Cf.* E. Wohlwill, *Galileo Galilei und sein Kampf fur die kopernicanische Lehre,* 2 v., Hamburg-Leipzig, 1909–1926; and my *Etudes Galiléennes,* Paris, Hermann, 1939.

[19] *Cf. Dialogo,* 167. This, of course, is false, but Galileo believed it to be true.

[20] *Cf. Discorsi e dimostrazioni matematiche intorno a due nuove scienze, Opere,* Ed. Naz., 8: 298.

[21] It is Cavalieri who, in his *Specchio Ustorio* (Bologna, 1632), published for the first time the demonstration that the trajectory of a projectile will be a parabola. It seems, nevertheless, certain that, long before Cavalieri, Galileo had been the first to find it out, and that Cavalieri was well informed about it; *cf.* E. Wohlwill, Die Entdeckung des Beharrungsgesetzes, in *Zeitschrift für Völkerpsychologie* 15: 107 sq., 1884, and Die Entdeckung der Parabelform der Wurflinie, *Abhandlungen zur Geschichte der Mathematik,* Leipzig, 1899.

[22] *Dialogo,* 190; Salusbury translation, p. 144; *cf.* p. 178 of the Santillana edition. The text of the *Dialogo* being easily available, I will not quote it in the original.

SALV. If the right motion towards the centre of the Earth was uniform, the circular towards the East being also uniforme, you would see composed of them both a motion by a spiral line, of that kind with those defined by *Archimedes* in his book *De Spiralibus*, which are, when a point moveth uniformly upon a right line, whilest that line in the mean time turneth uniformly about one of its extreme points fixed, as the centre of his gyration.[23] But because the right motion of grave bodies falling is continually accelerated, it is necessary, that the line resulting of the composition of the two motions do go always receding with greater and greater proportion from the circumference of that circle, which the centre of the stones gravity would have designed, if it had always staid upon the Tower; it followeth of necessity that this recession at the first be but little, yea very small, yea, more, as small as can be imagined, seeing that the descending grave body departing from rest, that is, from the privation of motion, towards the bottom and entering into the right motion downwards, it must needs passe through all the degrees of tardity, that are betwixt rest, and any assigned velocity; the which degrees are infinite; as already hath been at large discoursed and proved.[24]

It being supposed therefore, that the progresse of the acceleration being after this manner, and it being moreover true, that the descending grave body goeth to terminate in the centre of the Earth, it is necessary that the line of its mixt motion be such, that it go continually receeding with greater and greater proportion from the top of the Tower, or to speak more properly, from the circumference of the circle described by the top of the Tower, by means of the Earth's conversion; but that such recessions be lesser and lesser *in infinitum;* by how much the moveable finds itself to be less and less removed from the first term, where it rested. Moreover it is necessary, that this line of the compounded motion do go to terminate in the centre of the Earth.[25] Now having presupposed these two things, I come to describe about the centre *A* [fig. 4] with the semidiameter *AB*, the circle *BI*, representing to me the Terrestrial Globe, and prolonging the semidiameter *AB* to *C*, I have described the height of the Tower *BC*; the which being carried about by the Earth along the circumference *BI*, describeth with its top the arch *CD*: Dividing, in the next place, the line *CA* in the middle at *E*; upon the centre *E*, at the distance *EC*, I describe the semicircle *CIA*: In which, I now affirm, that it is very probable that a stone falling from the top of the Tower *C*,

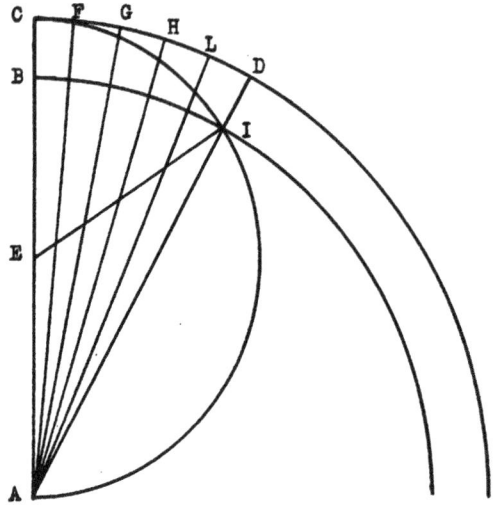

FIG. 4

doth move, with a motion mixt of the circular, which is common, and of its peculiar right motion. If therefore in the circumference *CD*, certain equal parts *CF, FG, GH, HL*, be marked, and from the points *F, G, H, L* right lines be drawn towards the centre *A*, the parts of them intercepted between the two circumferences *CD* and *BI*, shall represent unto us the same Tower *CB*, transported by the Terrestrial Globe towards *DI*; in which lines the points where they come to be intersected by the arch of the semicircle *CI*, are the places by which from time to time the falling stone doth passe; which points go continually with greater and greater proportion receeding from the top of the Tower. And this is the cause why the right motion made along the side of the Tower appeareth to us more and more accelerate. It appeareth also, how by reason of the infinite acutenesse of the contact of these two circles *DC*, *CI*, the recession of the cadent moveable from the circumference *CFD*; namely from the top of the Tower, is towards the beginning extream small, which is as much as if one said its motion downwards is very slow, and more and more slow *in infinitum*, according to its vicinity to the term *C*, that is to the state of rest. And lastly it is seen how in the end this same motion goeth to terminate in the centre of the Earth *A*.[26]

SAGR. I understand all this very well, nor can I perswade myself that the falling moveable doeth describe with the centre of its gravity any other line, but such an one as this.

SALV. But stay a little *Sagredus*, for I am to acquaint you also with three Observations of mine, that its possible will not displease you. The first of which is, that if we do well consider, the moveable moveth not really with any

---

[23] The reasoning of Salviati is completely erroneous; he reasons as a mathematician and not as a physicist. It is very significant that, with the exception of Borelli (and of Riccioli), this kind of reasoning will be followed throughout the whole discussion.

[24] The problem whether a body, starting from rest, passes all (an infinite number of) the degrees of velocity or tardity, i.e., the problem of the continuity or discontinuity of the acceleration, plays an important part in the discussions of the seventeenth century. It is well known that, while Galileo asserts the continuity *in all cases,* Descartes believes in the possibility, and even the normality, of a discontinuous "jump" from rest to motion. This is by no means a futile problem: the Galilean solution implies, indeed, not only the puzzle of the infinitesimals, but also the negation of hardness as a possible attribute of matter.

[25] The belief, shared by all participants in the discussion, that falling bodies must, in any case, reach the center of the Earth, is one of the main factors which prevent the correct solution of the problem. This belief inspired even Newton. It was R. Hooke who, for the first time, recognized clearly that, on a moving (rotating) Earth, the falling body will not reach its center; *cf.* my paper quoted *supra,* note 1.

[26] It is difficult to understand why Galileo did not follow his own inspiration and why, having stated that a uniform centripetal motion would produce a spiral (Archimedean), he did not conclude that an accelerated centripetal motion would produce a spiral of a higher degree. This failure is made the more strange by the fact that he was later (1637) to tell Carcavi that he had considered such a spiral (of the 2nd degree) "a long time ago," that is, in any case, long before the publication of the *Dialogo; cf. infra,* note 68.

more than onely one motion simply circular, as when being placed upon the Tower, it moved with one single and circular motion. The second is yet more pleasant; for, it moveth neither more nor lesse than if it had staid continually upon the Tower, being that to the arches CF, FG, GH, etc. that it would have passed continuing always upon the Tower, the arches of the circumference CI are exactly equal, answering under the same CF, FG, GH etc. Whence followeth the third wonder, that the true and real motion of the stone is never accelerated,[27] but always even and uniforme, since that all the equal arches noted in the circumference CI, are past in equal times; so that we are left at liberty to seek new causes of acceleration, or of other motions, seeing that the moveable, as well standing upon the Tower, as descending thence, always moveth in the same fashion, that is, circularly, with the same velocity, and with the same uniformity. Now tell me what you think of this my fantastical conjecture.[28]

SAGR. I must tell you, that I cannot with words sufficiently expresse how admirable it seemeth to me; and for what at present offereth itself to my understanding, I cannot think that the business happeneth otherwise; and would to God that all the demonstrations of Philosophers were but half so probable as this. However for my perfect satisfaction I would gladly hear how you prove those arches to be equal.

SALV. The demonstration is most easie. Suppose to yourself a line drawn from I to E. And the Semidiameter of the circle CD, that is, the line CA, being double the semidiameter CE of the circle CI, the circumference shall be double to the circumference, and every arch of the greater circle, double to every line arch of the lesser; and consequently, the half of the arch of the greater circle equal to the whole arch of the lesser. And because the angle CEI made in the centre E of the lesser circle, and which insisteth upon the arch CI, is double the angle CAD, made in the centre A of the greater circle, to which the arch CD subtendeth; therefore the arch CD is half of the arch of the greater circle like to the arch CI, and therefore the two arches CD, and CI are equal; and in the same manner we may demonstrate of all their parts. But that the business, as to the motion of descending grave bodies, proceedeth exactly thus, I will not at this time affirm; but this I will say, that if the line described by the cadent moveable be not exactly the same with this, it doeth extream nearly resemble the same.

SAGR. But I, Salviatus, am just now considering another particular very admirable; and this it is; That admitting these considerations, the right motion doeth go wholly wanting, and that Nature never makes use thereof, since that, even that use, which was from the beginning granted to it, which was of reducing the parts of integral bodies to their place, when they were separated from their whole, and therefore constituted in a depraved disposition, is taken from it, and assigned to the circular motion.[29]

---

[27] According to Riccioli (cf. infra, p. 352 and n. 116) it was just this possibility of explaining, or explaining away, the acceleration of the falling body that induced Galileo to adopt the circular solution. Indeed, the translation of Salusbury is not quite exact and does not render the true meaning of the text. Galileo says, as a matter of fact: "Talche noi venghiamo liberi di recercar nuove cause di accelerazione o di altre moti," etc., which means that we are freed from the necessity of probing for the causes of acceleration (because it does not exist) and not, as Salusbury translates it, that we are free to do so.

[28] "Questa mia bizzarria."

[29] This rather strange and "un-Galilean" assertion corresponds to others, just as strange, which we find at the very beginning of the Dialogo (Opere 7: 43, 56, 156) and according to which moto retto impossibile esser nel mondo ben ordinato and moto

SALV. This would necessarily follow, if it were concluded that the Terrestrial Globe moveth circularly; a thing which I pretend not to be done, but have only hitherto attempted, as I shall still, to examine the strength of those reasons, which have been alledged by the Philosophers to prove the immobility of the Earth, of which this first taken from things falling perpendicularly, has begat their doubts, that have been mentioned; which I know not of what force they may have been to Simplicius; and therefore before I passe to the examination of the remaining arguments, it would be convenient that he produce what he hath reply to the contrary.

III. GALILEO, MERSENNE, FERMAT

The Galilean solution of the problem of the trajectory of the falling body is, as we have seen, extremely ingenious and elegant. Unfortunately, it is quite false. It is even so obviously false (and besides incompatible with his own theory of uniformly accelerated motion of falling bodies) that one may wonder that Galileo did not see it himself.[30]

It is not impossible, as a matter of fact, that he did not persist in his error. This would explain why, in the Discorsi, he does not say a word about the "semicircular" theory of the fall. Indeed, at the very place where it should be discussed, that is, when, having demonstrated that the path of a projectile will be a parabola, Salviati has to meet the objections both of Sagredo and of Simplicio, he points out that this theorem, though correct in abstracto, will not describe the real motion of the projectile on the Earth, and merely concedes that the parabolic figure of the trajectory "would be greatly altered if it should terminate at the center of the Earth."

It is, of course, possible, too, that Galileo abstained from mentioning his "semicircular" theory for the very simple, and very sufficient, reason that it was linked together with that of the diurnal motion of the Earth. A subject that he had cause enough to avoid.

Yet be this as it may. The passage I am referring to is, nevertheless, extremely interesting as it seems to indicate, or at least to hint at, the motive which inspired Galileo in adopting the "semicircular" theory of fall, that is the necessity to incurve the parabolic trajectory in order to let falling bodies reach the center of the Earth.

Indeed, says Sagredo, after having heard Salviati's demonstration that the curve described by the projectile will be a parabola,[31]

---

retto assegnato a i corpi, naturali per ridursi al ordine perfetto quando ne siano rimossi; cf. my Etudes Galiléennes 3: 49 sq.

[30] As Mersenne will point out (cf. infra, p. 338), a body falling to the center of the Earth and describing in its fall a (semi)circular line would accelerate its motion not according to the Galilean law—space traversed proportional to the square of time—but according to a quite different one—space traversed proportional to the versine of the angle of rotation, this angle being itself proportional to the time.

[31] Discorsi (Opere 8: 275 sq.); I quote from the translation of H. Crew and A. de Salvio, Dialogues Concerning Two New Sciences, 250 sq., Evanston and Chicago, 1939.

SAGR. One cannot deny that the argument is new, subtle and conclusive, resting as it does upon this hypothesis, namely, that the horizontal motion remains uniform, that the vertical motion continues to be accelerated downwards in proportion to the square of the time, and that such motions and velocities as these combine without altering, disturbing, or hindering each other, so that as the motion proceeds the path of the projectile does not change into a different curve: but this, in my opinion, is impossible. For the axis of the parabola along which we imagine the natural motion of a falling body to take place stands perpendicular to a horizontal surface and ends at the center of the earth; and since the parabola deviates more and more from its axis no projectile can ever reach the center of the earth or, if it does, as seems necessary, then the path of the projectile must transform itself into some other curve very different from the parabola.

SIMP. To these difficulties, I may add others. One of these is that we suppose the horizontal plane, which slopes neither up nor down, to be represented by a straight line as if each point on this line were equally distant from the center, which is not the case; for as one starts from the middle [of the line] and goes toward either end, he departs farther and farther from the center [of the earth] and is therefore constantly going uphill. Whence it follows that the motion cannot remain uniform through any distance whatever, but must continually diminish. Besides, I do not see how it is possible to avoid the resistance of the medium which must destroy the uniformity of the horizontal motion and change the law of acceleration of falling bodies. These various difficulties render it highly improbable that a result derived from such unreliable hypotheses should hold true in practice.

SALV. All these difficulties and objections which you urge are so well founded that it is impossible to remove them; and, as for me, I am ready to admit them all, which indeed I think our Author would also do. I grant that these conclusions proved in the abstract will be different when applied in the concrete and will be fallacious to this extent, that neither will the horizontal motion be uniform nor the natural acceleration be in the ratio assumed, nor the path of the projectile a parabola, etc. But, on the other hand, I ask you not to begrudge our Author that which other eminent men have assumed even if not strictly true. The authority of Archimedes alone will satisfy everybody. In his Mechanics and in his first quadrature of the parabola he takes for granted that the beam of a balance or steelyard is a straight line, every point of which is equidistant from the common center of all heavy bodies, and that the cords by which heavy bodies are suspended are parallel to each other.

Some consider this assumption permissible because, in practice, our instruments and the distances involved are so small in comparison with the enormous distance from the center of the earth that we may consider a minute of arc on a great circle as a straight line, and may regard the perpendiculars let fall from its two extremities as parallel. For if in actual practice one had to consider such small quantities, it would be necessary first of all to criticize the architects who presume, by use of a plumb line, to erect high towers with parallel sides. I may add that, in all their discussions, Archimedes and the others considered themselves as located at an infinite distance from the center of the earth, in which case their assumptions were not false, and therefore their conclusions were absolutely correct. When we wish to apply our proven conclusions to distances which, though finite, are very large, it is necessary for us to infer, on the basis of demonstrated truth, what correction is to be made for the fact that our distance from the center of the earth is not really infinite, but

merely very great in comparison with the small dimensions of our apparatus. The largest of these will be the range of our projectiles—and even here we need consider only the artillery—which, however great, will never exceed four of those miles of which as many thousand separate us from the center of the earth; and since these paths terminate upon the surface of the earth only very slight changes can take place in their parabolic figures which, it is conceded, would be greatly altered if they terminated at the center of the earth.

In any case, even if Galileo himself did not recognize the falsehood of the semi-circular theory, others did very soon. Thus, in the summer of 1635, only three years after the publication of the *Dialogue on the two greatest world systems*, F. Mersenne, who made himself the *porte-parole* of Galileo in France [32]—he published the *Mécaniques de Galilée* in 1639—sent to Fermat (and others) a letter in which he pointed out some difficulties" of Galileo's theory: [33]

Propositions extracted from the Dialogue of Galileo among some others where some difficulties are to be found:

On pages 158 and 159 [34] he says [1] that it is fairly probable that a stone, falling from the summit of a tower to the center of the Earth, describes a semicircle, wherefrom it follows that falling moving bodies do not describe another line than a simply circular one; [2] that the falling body does not move faster than if it had remained at the summit of the tower; [3] that the motion of this moving body does not increase in falling, but remains uniform, as if it did not stir from its place.

And on page 160 he says that he does not want to assert that the motion of heavy bodies toward the Earth occurs precisely in this fashion, but only that if the line described by the falling body is not exactly this one, it approximates it very closely. Here Galileo was also largely mistaken, delighted, as we may suppose, by the beauty of the consequences that he draws from his propositions. For it is easy to see, as well by his figure on page 159,[35] as by the continuation of his discourse and the explanation of his figure that the moving body, passing through the diameter CIA, traverses it in six hours, because it is in this same time that the point of the tower C, from which the moving body departed, describes a quarter of the circle by the diurnal motion. And because Galileo does not determine the height of the Tower, and does not take into account the diameter of the Earth, it would follow from this proposition that, whatever the height of the Tower may be,

[32] Mersenne has usually been represented as the *porte-parole* and agent of Descartes; that this is by no means the case has been shown by R. P. R. Lenoble in his very important book, *Mersenne ou la naissance du Mécanisme*, Paris, J. Vrin, 1943.

[33] This letter has been published, though without being attributed to Mersenne, by A. Favaro in the *Atti e Memorie della Reale Accademia Venete in Padova*, Anno CCXCVI, Nuova serie, 2: 34–35, 1894–1895, and has been republished by Cornelis de Waard in his paper, La spirale de Galilée, in the supplementary volume of the *Œuvres de Fermat*, 10 sq., Paris, Gautiers-Villars, 1922. It is Cornelis de Waard who identified the author of this letter with Mersenne. Mersenne's text has been translated from the French. I did not improve the style of Mersenne which may be judged by the passages I am quoting in French.

[34] Mersenne quotes, of course, the original edition of the *Dialogo*.

[35] I am not reproducing the figure which is, of course, that of p. 334.

and should it reach to the Moon or the Sun, or even farther, the moving body would always use the same time for descending to the center; and if the Earth should be no larger than ☿ [Mercury], or should have only a hundred leagues for diameter, or less, the moving body would not take less time for passing from the surface of the body to the center. Which is not credible, and Galileo does not give any proof [of it].

But if we suppose the experiences concerning the space that heavy bodies traverse in falling, and that the spaces traversed are in a duplicate ratio to times,[36] as he assures us to have discovered on pages 17 and 217, we will find only 20 minutes of an hour (somewhat less) for the time that a cannon-ball would take for descending to the center of the Earth, during which time the Earth makes but 5°, which is rather far from 90°. And because the observations of Galileo do not agree with ours, and because he makes this motion somewhat slower, the time of the fall of the moving body would be more than 25½′ [minutes], according to his observations, while the Earth would make 6° 22′. Wherefrom it would follow that the line described by the moving body will be very different from the semicircle, and that it would be rather notably curved near the circumference, but approaching the center, it would be difficult to distinguish it by sight from the straight line.

Now Galileo draws from [his theory] another consequence which is that nature does not use straight lines at all for reuniting parts separated from their wholes, but only circular ones. But, even if we should receive this proposition as true, this circular line would be described only on the equator, considering only the diurnal motion; for, if we should combine it with the annual, it would be very far from truth that the line should be circular, in whatever place of the Earth the moving body were located. Positing only the diurnal [motion], I say that below the poles heavy bodies fall in a straight line, which consequently, would not be entirely banished from nature; and on the parallels, they would describe curved lines which would approximate the circular all the more so as they would be nearer to the equator.

Father Mersenne seems to have been rather proud of his discovery. In any case, being the last man to hide his light under a bushel,[37] he proceeds immediately to announce it to the world: first in Latin in his *Harmonicum libri sex* (in 1635), then in the French *Harmonie Universelle* (likewise in 1635[38]—Mersenne was in the habit of publishing his works both in Latin and in French), and finally once more in Latin in his *Cogitata Physico-Mathematica* (1644). Each time he gives a very careful account of Galileo's theory and even an improved one with a diagram of the trajectory much more elaborate than that of Galileo—he tells us that he had formed the conception of the circular fall independently, before having read the *Dialogue*[39]—followed

by a criticism which leads to his own determination of the trajectory of the falling body. In the last named publication his own solution is left out. Instead Mersenne gives an account of the theory developed in the meanwhile by Fermat which we may assume he accepted as true.

Let us now examine these texts, or at least the much more elaborate French one of the *Harmonie Universelle*. Mersenne begins by giving a report of Galileo's criticism of Scheiner's (Locher's) objections to the diurnal motion of the Earth: [40]

Although many people believe that stones and heavy bodies fall all the more quickly toward the center of the Earth when they are heavier, nevertheless, experience has shown the contrary, as I will show, after having mentioned the experiences of Galileo which he uses in order to refute the book of *Mathematical Conclusions* of Schener [sic], where it is objected against the diurnal motion of the Earth that it would imply that an artillery bullet carried by an Angel up to the concave of the Moon, would use more than six days for falling to the Earth, even if its motion were as quick as the circular [motion] of the great orb of the moon, namely, that it should traverse 12.000 German miles in every hour; that it is unbelievable that it should always remain at the vertical point during the six days that it would turn around with the Earth, describing under the Equinoxial [41] a spiral line in the plane of the great circle, under the parallels a spiral around the cones, and a straight line under the poles.[42]

To this Galileo replies that, as the semidiameter of the circle is smaller than the sixth part of the circumference, it follows that the bullet, having only the semidiameter of the [orb of the] Moon to descend, will be on the Earth long before the heaven of the Moon will make the 6th part, as he supposes that the bullet is as fast as the heaven of the Moon; and that it will fall in less than 4 hours, supposing that the said heaven makes its circuit in 24 hours, which one has to suppose in order to make the weight remain on the same vertical line.

Yet it seems to Mersenne—and he is perfectly right—that Galileo's criticism is not consistent with his own views about the fall of heavy bodies towards the center of the Earth, and he continues by pointing it out:

But it seems that Galileo did not think here about the the circular motion of the bullet which would fall only in six hours, supposing that it would have the same speed as the heaven of the Moon and that its speed would not increase in any way because of its approximation to the Earth, as Schener supposes it; for in this case the bullet would fall along a semicircle equal to the quarter of the circle of the Heaven of the Moon, as I shall demonstrate in a short while.

This demonstration, which differs from that of Galileo and tries to improve it, is rather curious. It occupies,

---

[36] "En raison doublée des temps" = proportional to the square of the time.

[37] Father Mersenne has been rightly, though somewhat nastily, called the "boîte aux lettres" of the scientific world of his time.

[38] The printing of the immense volume (1486 pages *in folio*) took a lot of time, and parts of it were already circulating by the autumn of 1635.

[39] Cf. *infra*, p. 338. A number of people seem to have arrived at the discoveries—and even at the errors—of Galileo without having read his works. At least they pretended to have done so, unfortunately without proof.

[40] Cf. *Harmonie Universelle contenant la théorie et la pratique de la Musique, ou il est traité de la Nature des Sons et des Mouvemens, des Consonances, des Dissonances, des Genres, des Modes, de la Composition, des Voix, des Chants, et de toutes sortes d'Instruments harmoniques*, par F. Marin Mersenne, de l'Ordre des Minimes, à Paris, chez Sebastien Cramoisy, 90 sq., 1636.

[41] "L'Equinoxial" = the equator.

[42] It is obvious that Mersenne had not read Locher's *Disquisitiones*, but only the report made on them by Galileo.

together with its refutation which shows that the motion on a semicircle is incompatible with Galileo's law of fall, the whole of the propositions III and IV of book II of the *Harmonie Universelle:* [43]

To determine the figure [trajectory] of the motion of heavy bodies which would fall from the summit of a Tower, or from whichever height one should want, supposing that the earth moves and makes every day a full revolution on its axis.

It is not necessary to explain the helix [spiral] that the weight would make if its motion were uniform, like that of the Earth, as we have shown its difformity and its inequality,[44] which I now assume, to avoid repetition. But in order that this proposition be more agreeable, I shall examine the thoughts of Galileo on this subject, of which he speaks starting on page 156 of his *Dialogue*; accordingly I describe here the circle *BI* [fig. 5] from the center *A*, which represents the Earth and produce the semidiameter *AB* until *C*, in order that *BC* be the height of the tower which, being carried by the Earth on the circumference *BI*, describes by its summit the arc *CDN*. I divide thereafter the semidiameter *AC* by the middle at point *E*, wherefrom I describe the semicircle *CIA*, along which circle, according to Galileo, it is probable that the stone falls if its motion is composed of the circular motion of the Earth, and of the straight which is proper to it, which he proves as follows:

If in the circumference *CD* we mark several equal parts, such as *CF*, *FG*, *GH*, *HL*, and *LD*, and from the points *F*, *G*, *H*, *L*, we draw perpendiculars to the center *A*, the parts of the lines comprised between the two circumferences *CDN* and *BIM* represent the tower carried by the Earth from *C* to *N*, and the points where the diameters cut these lines will be the places where the stone will, from time to time, find itself in falling. Now these points withdraw ever more and more from the summit of the Tower; that is why the straight motion of the stone along the Tower appears even more increased and more violent. And because the angle *DCI* is infinitely acute the withdrawal from the surface *CFD*, or from the summit of the Tower is very small at the beginning, and consequently the motion of the stone is all the slower as it is nearer to *C*, or to rest, and it moves quicker close to the center *A* than in any other place.

Thus the Galilean solution seems, at least at first glance, to give a perfect explanation of the phenomenon of free fall. Alas, as Mersenne will carefully show, it leads to rather unacceptable consequences and is rigorously incompatible with Galileo's own law of fall according to which the distance traversed by the free-falling body is proportional to the square of the time. Indeed, if bodies fell along the circumference of a circle they would conform to a quite different proportion (that he noticed it, gives us the measure of Mersenne's ability), namely that of the arcs of the circle to their versines. Of course, we cannot decide experimentally between the two proportions—Mersenne, once more, is perfectly right—yet, if the latter were true, all bodies, whatever their distance from the center of the Earth,

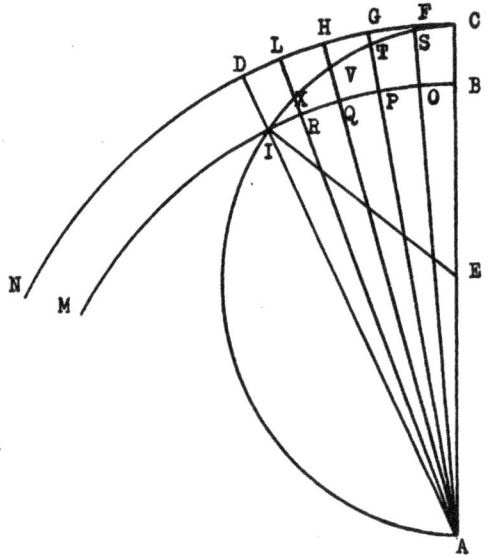

FIG. 5

would reach it at the same time, which means that they would move with speeds proportional to this same distance. This results in a state of affairs which appears to Mersenne so improbable that it practically amounts to an impossibility, and thus refutes itself. Thus writes Mersenne:

We must examine this beautiful thought of Galileo, in order to see whether the motion of the stone, which seems to us perpendicular, can be circular and equal to that of the earth, as the semicircle *BIA* is equal to the quarter of the circle *CN*. We have already considered this line before having seen his *Dialogue*; but as he puts the fall of bodies in double proportion to the times, as we have done,[45] to which the proportion of the versines of equal arcs is *quasi*-similar,[46] especially at the beginning of the fall, when they are small, it is easy to show that the fall of stones cannot be performed along the semicircle *BIA*; which I demonstrate by the other figure, which follows, namely, *A*, 90, *L* [fig. 6], in which the arcs represent the time, and the versines the space of the fall; that is why, when the place from which the weight falls, that is *A*, will be carried by the diurnal motion as far as 9, which means 9 degrees, which will be accomplished in 36' of the hour, the weight will be in *B* and, accordingly, will be at point 2 [1] of the semicircle,[47] that is, the place where the perpendicular 9 *L* cuts the arc described from point *B*; and when it will be at point 18, which will be the case in an

---

[43] *Cf. Harmonie Universelle,* l. 2: 93 sq.

[44] It is interesting to note the use of scholastic terminology—we shall see it used by Riccioli as well—for the expression of Galilean ideas. Mersenne means that he has already established the motion of the fall to be a uniformly accelerated one (*cf. Harmonie Universelle,* 88 sq.).

[45] The expression used by Mersenne is rather curious; it suggests—without openly asserting it—the independent discovery of the law of the fall. Mersenne should have said: "puisqu'il a établi la raison doublée des temps que nous avons acceptée de lui."

[46] For altitudes which are practicable for experimental control, the two formulas give nearly equivalent results in the beginning of the fall.

[47] Mersenne writes 2, but it should be 1. I have corrected his drawing, or, more exactly, his notation of the points.

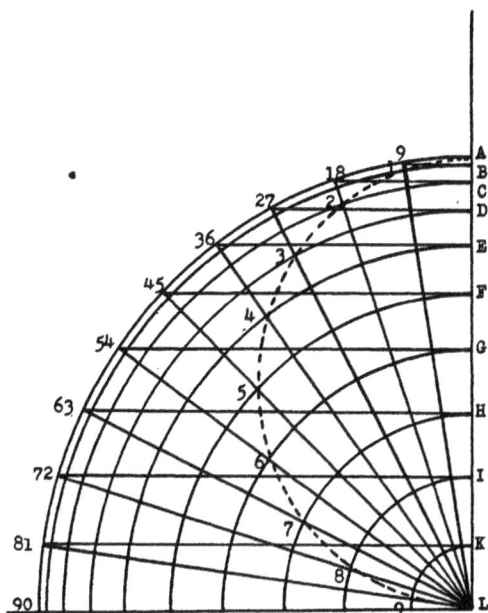

FIG. 6

cause it suffers very little violence, and does not desire to change its place with as much violence as when it is removed farther away from its center.[49]

For we can reply the same thing concerning the weights removed from the earth and say that they always come back to this center in an equal time, as does the lead tied to a rope, about which we shall speak later on; and we can even add that the equality of the returns to the center of the earth is more rigorous than that of the return of the lead,[50] because the cord impedes it a little, whereas the stone which falls straight to the center does not have this impediment, and this time of the fall of the stone which falls straight to the center would always be six hours, according to the preceding hypotheses, even if it should fall only from a height of a foot from the center.

In any case, it is impossible to make about it experiments which would convince the contrary, all the more so that as whatever heights one should take, the difference of the velocities will be so small that no human industry can perceive it, any more than the difference between the double ratio of the spaces to the times of the fall and that of the versines of the double arcs;[51] for the radius being 10,000 times the versine [of an angle] of 15′, which arc traversed in 1′ [minute of time] by the diurnal motion, is one, that of 30′ is 4, and that of 45′ is 9. But we are not able to observe here a fall in 1′ of an hour, and even less in 2′, and nevertheless, at this distance, the ratio of the spaces is exactly double that of the times, even if it should be the ratio of the versines to the arcs. But if we should proceed further, or if we were nearer to the center, we should find a manifest difference, and the ratio of the versines to the arcs will always be smaller than the double ratio of the spaces to the times; although we cannot know the true proportion that the weight would keep till the center, and we could maintain, therefore, that it is that of the versines of the arcs. Nevertheless, as the double ratio

hour and 12′, the stone will be at point 2, that is in the place where the arc coming from C cuts the perpendicular 18 L, for the space 18, 2 is equal to the space of the fall AC, and to the versine of the arc A 18, which represents the time.

Similarly, when the point A is carried as far as 27, the weight will be as far as D, from where the quarter of the circle being described, it meets the perpendicular 27 L at point 3. When A will be in 45, which will take place in 3 hours, the stone will be at point 5, which is the place where the arc described from point F (which is the endpoint of the fall of the weight F because it is the versine of 45 degrees) cuts the perpendicular 45 L. And consequently, the weight arriving at 72 in 4 hours 48′, the versine of 72 degrees is in AI, and the arc described from point I meets the perpendicular 72 L at point 9 [8],[48] which is the place of the weight; so that the point A being in 90, the weight will arrive in 6 hours at the center of the Earth L along the semidiameter A, 1, 2, 3, 4, 5, 6, 7, 8, 9, L: Wherefrom it follows that all kinds of weights, of sufficient matter and gravity for overcoming [the resistance] of the air, such as lead, stones, wood, and so on, must fall to the center of the Earth in six hours from whatever distance [they move], for instance from a league near the center as well as from the Moon; which is impossible if the weights begin their motion with the same speed in all kinds of places and if they maintain continuously the proportion we have spoken about; so that in order to verify the preceding fall it would be necessary for the weight to descend all the more slowly as it falls from a height less removed from its center, as in the case of lead suspended by a long cord, of which the motion is all but imperceptible when it is very little removed from its line of direction [the perpendicular], be-

[48] Once again Mersenne makes an error; he writes 9 in place of 8.

[49] The interesting hypothesis that bodies are the more strongly drawn toward the Earth's center the farther they are from it is not an invention of Mersenne, but a theory that has actually been propounded by the so-called "géostaticiens"; it is, by the way, an inevitable consequence of the explanation of gravity by attraction for bodies that are inside the Earth. Cf. Cornelis de Waard, La chute des graves, in *Œuvres de Fermat*, suppl., 20 sq.

[50] The lack of exact isochronism in the oscillations of a pendulum is thus observed by Mersenne as early as 1635. He deals with the problem in his *Cogitata Physico-Mathematica* of 1644.

[51] Mersenne, as we know, is rather sceptical about the possibility of achieving an exact science: neither reasoning nor experience can lead us to this goal. Cf. *Harmonie Universelle*, 112, Corollaire II: "Ceux qui ont veu nos experiences, et qui y ont aidé, scavent que l'on n'y peut proceder avec plus de iustesse, soit pour le plan qui est bien droit, et bien poli, et qui contraint le mobile de descendre droit, ou pour la rondeur, et la pesanteur des boulets, et pour les cheutes: d'ou l'on peut conclure que l'experience n'est pas capable d'engendrer une science, et qu'il ne se faut pas trop fier au seul raisonnement, puisqu'il ne respond pas toujours à la vérité des apparences dont il s'eloigne bien souvent: ce qui n'empeschera pas que je ne parle du plan également incliné, tel qu'il doit estre afin que les corps pesans le pressent et pesent egalement sur chacun de ses points. Si quelqu'un desire faire les experiences plus iustes, il doit user d'un plan incliné plus long que le nostre; par exemple d'un plan de 48 pieds sur lequel le temps de la cheute sera beaucoup plus sensible: et si l'on en avoit un de cent ou deux cent pieds, il seroit encore meilleur." Cf. my paper, An experiment in measurement, *Proc. Amer. Philos. Soc.* 97 (2): 222–237, 1953.

is easier, it is preferable to use it instead of the other, as it cannot mislead us on the surface of the earth.[52]

But before proceeding further, it is worth noticing that the reason which made us consider this fall along the semicircle is this: having assumed the diurnal motion of the Earth, and also that the weight is carried from point A to point 18 along the perpendicular [line] 18 L, as near to the center L as the point C, which is touched by the line 18 C, parallel to the horizon of the place of the motion of the fall 90 L, and having only considered the fall along the perpendicular AL, in the proportion in which the arc A, 9, 18, 27, and so on, bends itself, so that having reached point 90, the line, which is drawn therefrom perpendicularly upon AL, terminates at the center of the Earth (but as the weight does not remain upon the line AL, because it follows the diurnal motion, when the place where the weight falls from arrives at point 27, after having drawn from this point a perpendicular on AL, which touches it at point D, and which shows the fall of the weight, while the Earth has made the arc A 27), I draw an arc from point D, the center L remaining always the same, in which case the place where the arc meets the perpendicular drawn from point 27 to the center, that is [point] 3, is the place of the weight, because it has found the arc D 3, and having marked several places equally distant upon the quarter of the circle A 90; and having drawn perpendicular lines cut by several other smaller quarters of the circle, according to the places where the first line AL is cut, and furthermore the line which marks the fall being drawn by the perpendiculars and by the arcs which cut each other, we have finally found that this line was a perfect semicircle, and that the arcs parallel to the quarter of the circle AC are removed from each other according to a proportion very near to the double, and so similar to the arcs removed from the center L, that the difference cannot be noticed by any observation.

Mersenne states—and this is very interesting—that unfortunately it is impossible to settle the question by observation and experience: the distances available for an experiment are much too small, and thus the differences (for instance the differences in the speed of the first moment of the fall) which cannot be deduced theoretically cannot be observed. Yet the law of fall can be considered as sufficiently well established to allow us to reject the theories the consequences of which contradict it. Thus he continues:

But after having examined this matter at leisure we have found that it was impossible, according to our experiences, and one or the other of the said proportions, that a weight should take six hours to descend from the surface of the Earth to the center, and also that our thought was not of a great consideration for demonstrating the fall of heavy bodies by the circular motion, that is for demonstrating that its speed increases when the arc comes nearer to the horizontal line 90 L; therefore the path of the fall on a semicircle which we have described by the means explained here, could not be defended, because it leads to great absurdities, that must be examined in proposition IV.

The absurdities which Mersenne has in mind do not depend on the law—or the proportion—relating the time of fall to the distance traversed, but to the experi-

mentally ascertainable *speed* of the downward motion. Thus they would occur in both cases, i.e., if bodies fell according to "doubled ratio of times" or if they did it according to the ratio of the versines to arcs.

Mersenne does not state it explicitly, but it is clear from his reasoning that he does not link together the speed of the descent according to the proportion of the versines to the arcs with that of the rotation of the Earth. He obviously considers it as an abstract pattern, quite independent of the actual value of the acceleration. The absurdity of the "semicircular" solution presented by Galileo is, therefore, not in the fact that it implies the isochronism of all the "falls"—this is bad enough, yet not absurd—but that, just because of the link with the earth's rotation, it postulates the six-hour period for the descent to the center of the Earth and thus contradicts his own (and Mersenne's) experiments.

The analysis of the "absurdities" which follow from the Galilean theory of circular fall, as well as the positive solution of the problem, the determination of the true trajectory of the falling body, is given by Mersenne in the following proposition IV of book II: [53]

To show that it is impossible that heavy bodies, descending to the center of the Earth, [should] describe the preceding semicircle, and to determine the line by which they would descend if the Earth turned in 24 hours around its axis.

We have already said that if the weight fell in six hours from whatever place it was, it would be necessary that it (the weight) should have different degrees of speed, according to the diverse distances from the center [of the place] where it is let to fall from, and consequently, the moon being at a distance of 58 terrestrial semidiameters or of 66,666 leagues from the center of the earth, it would traverse 820 leagues in 36′ [minutes] of the hour, because the earth would make 9 degrees in 36′. Now the versine of 9 degrees is 1,230, the radius being 10,000; accordingly, if the radius is 66,666, the versine will be 820 leagues, and if the radius is equal to the semidiameter of the earth, that is to 1,145½ leagues, the versine of 9 degrees will be 14 leagues, and therefore the weight should make only 14¹⁄₁₀ leagues in 36′; and nevertheless, according to our experiences, and the double ratio, it will fall 3,732 leagues and 1,200 fathoms during that time.[54]

This, of course, cannot be experimentally established. Yet there are some experimental data that are incompatible with the fall along the semicircle, as Galileo would have it:

The weight falls 108 feet [55] in 3″ [seconds of the hour] as it is always shown by very exact observations; and, nevertheless, if it should fall in six hours to the center, it would traverse only 4 inches and 11 lines, for in 3″ the Earth makes 45″ [seconds of the circle] of which the versine is 238, the radius being 10,000,000,000. And if the versine of the radius is 238, then 17,181,818 feet will give 4 inches 11 lines, which is a difference so remarkable that there is no place for doubting that a weight cannot take six hours to fall to the center. Yet, as it falls 108 feet instead of 4 inches 11 lines, then it should, according to this suppo-

---

[52] "On ne peut s'y méprendre." Mersenne wants to stress that the adoption of the Galilean proportion will not, for pragmatic reasons, lead to an error, or at least not to an observable one.

[53] *Harmonie Universelle*, l. 2: 96 sq.
[54] The *toise* (a fathom) is equal to six feet = 197.22 cm.
[55] The foot used by Mersenne is the *Pied du Roy* = 32.87 cm.

sition, be falling [during] 48½″ [seconds of the hour]; for if 17,181,818 feet give 108 for the versine, 10,000,000,000 would give 6,285⁷⁄₁₀, which being taken off from the radius, it remains 999,993,714 for the sine of the complementary [angle] that the earth makes during the fall, which is performed in 48½″ [seconds] which is a time too long and too different from 3″ as not to infer [therefrom] absurdities which follow from such an hypothesis; for the experience shows that it [the heavy body] traverses 263 times more space than it would do supposing the fall from the surface of the earth to the center in 6 hours, because 4 inches and 11 lines are so many times in 108 feet, and in order to traverse an equal space, it would use 16½ times more time that according to the double ratio, and to the experiment, as 3″ are as many times in 48½″ and this disproportion of time corresponds well to that of the spaces, for 1 to 263 is nearly in the double ratio of 1 to 16½.[56]

Thus it is easy to conclude from all this discourse that Galileo was satisfied with having a proportion of the fall which seemed to him to agree with the appearances, and that thinking more about the beautiful correspondences and consequences which he drew from it, he did not give a deep consideration to this matter, as it is not credible that such a man would be so mistaken if he had examined more thoroughly the fall of bodies according to the experiments that he made himself.

Yet, concludes Mersenne:

It is therefore evident that the moving body would not move in the same manner as if it remained in the place where it should fall from, that is, along a circular line, and, consequently, that it would not go as far as if it remained on the summit of the tower, and that it would not have a uniform and equal motion as Galileo imagined; for we have shown clearly that a weight cannot fall from the surface to the center in six hours, as it would be necessary, and that, according to our experiences and the double ratio, or that of the versines,[57] it will arrive at the center in 19′56½″ while the Earth will make 4 degrees 59⅛′, and if one follows the experience of Galileo, it will reach the center in 25½′ while the Earth will make 6 degrees 22⅔′.[58] Therefore, it will describe the curved line AB-DEFC [fig. 7], which is greatly different, not only from the semicircle, but from any part of the circle and of the arc one should design; for if one takes away the portion ABD, the rest is hardly different from a straight line as one sees it particularly in the portion EFC.

Now this line is described in the following manner: I draw the straight line AC which represents the semidiameter of the Earth, of which C is the center, and then I draw the line CO, which makes with AC an angle of 6 degrees 22⅔′; for, if the line AC is 1,000,000 the line AO will be 11,178. And then I divide the arc AO in five equal parts, of which each has one degree 16⅛′, and the line A [AC] in five unequal parts, of which the first has one [unit-part], the second 3, the third 5, the fourth 7, and

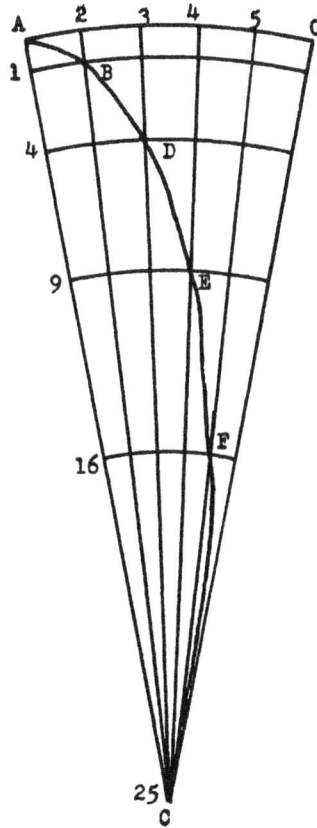

Fig. 7

the last 9, which makes together 25 [unit parts], that is the square of 5. And through these sections I draw arcs till the line OC, so that while the Earth turns and describes the arc A2, the weight falls as far as B in 5′ 6″; and describing the other arc 2, 3 in 5′ 6″, it falls from B to D, that is three times more; and then in the same time it traverses DE which contains 5 parts; and while the Earth makes the arc 4, 5, the weight falls the space EF, and then FC, and so on, increasing its speed in the double ratio of times.

If the weight fell from 373,248 leagues, that is from 326 terrestrial semidiameters, it would arrive in six hours at the center,[59] and the line of fall would describe a figure similar to the semicircle, supposing that the proportion be in the double ratio but if it were as the versines to the arcs, it would describe a perfect semicircle, and falling from another distance, it would describe a helix, if the distance were greater than 326 semidiameters. Which is easy to demonstrate as we have already done elsewhere. And we can also see several observations that I have made

---

[56] Thus it is not the abstract pattern of the semicircular relation (versines to radius) which is refuted by the experiments performed by Mersenne—this, as we know, is impossible—but the concrete utilization of this pattern by Galileo: the six-hour period.

[57] Once more, Mersenne believes that his (or Galileo's) abstract patterns and laws of fall are experimentally indistinguishable.

[58] "L'experience de Galilée" refers to the numerical data—the value of g—given by Galileo in the Dialogo (the distance traversed in the first second of the fall as equal to 5 cubits), data which Mersenne found to be extremely faulty; cf. Harmonie Universelle, 85, 87, cf. my paper quoted, n. 51.

[59] Mersenne, of course, assumes—as Galileo had before him, and Riccioli and even Borelli were to do after him—that the value of g is a world-constant.

upon this subject in the book *De Causis sonorum* [60] in the 24th and the 27th proposition.

Corollary

We can conclude from this proposition that all the thoughts and the experiences of Galileo do not favor the diurnal motion of the Earth.[61] And that the weights would never fall along a semicircle, even from the distance that we have supposed, except under the Equator, and that they would fall only along a straight line under the Poles.

We now shall turn our attention to Fermat's solution of the Galilean problem, a solution much more interesting than that of Mersenne, though of course just as false. It is certainly Mersenne's letter (quoted *supra* p. 336) which induced Fermat to think about the trajectory of the falling body and to reject both Galileo's and Mersenne's attempt to discover it. Fermat's own treatment of the problem is purely mathematical, and, as is to be expected, most elegant, and even brilliant. Yet, as a matter of fact, though following the inspiration of Galileo or perhaps just because of following it too strictly,[62] Fermat does not deal with the problem at all. He does not understand it because he does not understand the very basis of the new Galilean mechanics: the principle of inertia.[63] Therefore, he does not think of a body moving freely in space and animated at the same time by a tangential and a centripetal motion; he thinks of a body moving down, naturally, towards the center of the earth,[64] along a certain fixed radius and sharing at the same time the motion of this radius which turns around the center of the earth. He finds out, quite correctly, that the resulting line will be a spiral.

Thus he writes to Carcavi sometime between the third of June, 1636, and the beginning of 1637 that as Galileo (a man of great genius) has in his *Dialogue* tentatively asserted that a heavy body in its natural downward fall—the diurnal motion of the earth being supposed—would describe a semicircle; and as this opinion has already been proven false,[65] he, Fermat, undertook to investigate more closely the true nature of this line, which he believes to be identical with the line

[60] The *De causis sonorum* is a part of Mersenne's Latin *Harmonicorum libri six*, Paris, 1636.

[61] Although Mersenne asserts Galileo's theory to be unfavorable to the motion of the earth, he does not use his criticism of Galileo himself as an argument against that motion. On the contrary, his own theory is expressly based upon the supposition of the earth's diurnal motion.

[62] Fermat, who obviously follows the lead given by Galileo (*cf. supra* p. 334 and n. 26), adopts also his mistaken starting-point, and, not being troubled by any physical scruples—certainly the case with Galileo—he develops the implications of the initial error to their logical end—something which Galileo did not do.

[63] Of course one could, and should, say that Galileo himself does not understand—at least not clearly—his own problems.

[64] Fermat, like everybody else, is convinced that falling bodies strive to reach the center of the earth.

[65] Fermat, beyond doubt, has in mind the refutation of Galileo by Mersenne.

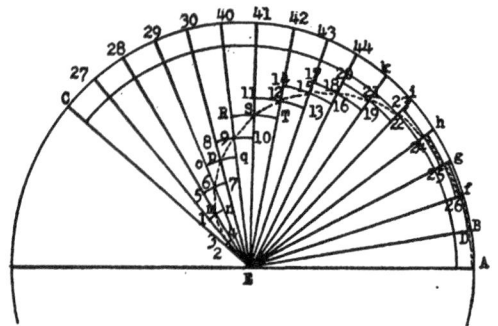

Fig. 8

which Menelaus, as reported by Pappus, calls admirable.[66]   Indeed, explains Fermat,[67]

Let there be on the Earth the circle *ABKC* [fig. 8] which we will assume is its equator, and let the beginning of the motion of a certain heavy body be in *A*, and let the downward fall to the center *E* take an amount of time which to the time of 24 hours has the same relation as the arc *AKC* to the entire circumference (supposing only the diurnal motion of the earth and not the yearly one), I say that the falling body will describe not a semicircle, but a certain *spiral*, such as is *A, D*, 21, 18, 15, 12, etc., which

[66] The passage alluded to by Fermat is in the Prop. XXX of the Problem VII of Book IV of the *Collectiones Mathematicae* of Pappus.

[67] Lettre de Fermat à Carcavi, in *Œuvres de Fermat* (La spirale de Galilée), suppl., 15 sq. The solution of Fermat having played a very important part in the history we are pursuing, I shall quote it in the original: "Cum Galilaeus (magni ingenii vir.) in suis Dialogis dubitanter asseruisset graue (supposito diurno Telluris motu) naturaliter descendens, semicirculum descripturam, nobis occasio fuit attentius veritatem inquirere, jamque demonstrato huius opinionis errore veram lineam damus, quae, ni fallor, eadem est quae a Menelao apud Pappum mirabilis appellatur.

"Sit in Tellure circulus ABKC quem hic aequatorem supponimus, et sit gravis alicuius motus principium in *A* fiatque descensus usque ad centrum *E* in tempore quod ad tempus 24 horarum se habeat ut arcus *AKC* ad totam circuli circonferentiam (supposito diurno Telluris motu tantum, non autem annuo), dico a gravi descendente describi non semicirculum, sed *helicem* quamdam, qualis est *A, D,* 21, 18, 15, 12 etc., quae in figura punctis notatur, cuius (supposita proportione descensus gravium a Galileo assignata) haec est proprietas, *ut ducta a centro quaecumque recta EK, secante helicem in puncto 21 et circonferentiam circuli in K, ductis etiam rectis EA et EC, recta EK ad rectam K21 sit in ratione duplicata circonferentiae AKC ad circonferentiam AK.* Alia proprietas insignis haec est *ut spacium helice AD 21 E et recta EA contentum, ad sectorem AKCE sit ut numerus octo ad numerum quindecim.*

"Prima proprietas facile probatur ex natura ipsius helicis. Diuiso enim arcu *AKC* in quotlibet partes aequales, ductisque a centro ad puncta divisionum semidiametris, diviso etiam tempore descensus in tot partes aequales et posito quod linea *BD*, quam percurrit grave in prima parte temporis, sit unius mensurae, erit linea *f,* 26, quam in duobus primis temporibus percurrit grave, quattuor mensurarum; *g,* 25, novem mensurarum; *i,* 23 viginti quinque mensurarum; *K,* 21 triginta sex mensurarum; atque ita secundum naturalem ordinem numerorum quadratorum. Unde patet conclusio."

in the drawing is designated by points; which [spiral] (if we admit the law of descent of heavy bodies asserted by Galileo) has the property that if we draw a straight line EK, cutting the spiral in the point 21 and the circumference of the circle in K, and if we draw the straight lines EA and EC, the straight line EK will be to the straight line K 21 in the duplicate ratio of the circumference AKC to the circumference AK. Another very remarkable property is that the space contained between the spiral AD21E and the straight line EA will be to the sector AKCE as the number eight to the number fifteen.

The first property is easily proved from the very nature of the spiral. Indeed, the arc AKC being divided in whatever [number of] equal parts, and semidiameters being drawn from the center to the points of division, and the time of the descent being divided in the same number of equal parts, and assuming the line BD which the heavy body traverses in the first part of the time to be of one measure. then the line f, 26 which the body will traverse in the two first parts of the time will be of four measures; g, 25. of nine measures; i, 23, of twenty-five measures; K, 21, of thirty-six measures; and so forth according to the natural order of square numbers. From which the conclusion is evident.

It is interesting to note that Galileo, informed by Carcavi of Fermat's (and Mersenne's) criticism, acknowledged his error; or rather he acknowledged the falsity of the solution developed by him in the *Dialogue*, pretending at the same time that he never held it to be true and had presented it only as a "joke." More interesting still is the fact that he seems to have accepted that of Fermat. In any case, he does not criticize it, but only tells Carcavi (i.e., Fermat) that long ago he already had in mind the idea of such a curve, i.e., of a spiral of a higher degree than that of Archimedes.

Thus he writes to Carcavi [68] (for Fermat) that

it is a long time since, having seen and studied with extreme admiration the spiral of Archimedes which he compounded by two uniform motions, the one straight and the other circular. I formed in my mind the spiral built up by a uniform circular motion and a straight one, accelerated according to the ratio of acceleration of the heavy bodies in their natural descent, which I am convinced is in a duplicate proportion to that of the time; and that is the spiral of the friend of your Lordship.

And though it has been said in the *Dialogue* that it may be that the mixing together of the straight [movement] of the falling body with the uniform circular [movement] of the diurnal motion will form a semi-circumference which will terminate in the center of the Earth, this has been said in the manner of a joke, as it appears clearly enough, from its having been called a *caprice*, and a *whim*, that is *jocularis quaedam audacia*.

It is hard to believe that Galileo had really meant his solution of the trajectory of the falling body to be merely a joke—but be this as it may. It is more important, in my opinion, that, just like Mersenne and Fermat, he seems to build up this trajectory by combining a straight motion downwards (towards the center of the Earth) with a *uniform circular* (and not a tangential) one, confusing, moreover, just as his predecessors (and successors) did, the uniform circular motion of a body with that of its radius-vector; and that, just like Mersenne and Fermat and everyone else but Hooke (Newton included) he is deeply convinced that the body falling thus will finally arrive at the center of the Earth.[69] This shows us how difficult it was, even for its creators, to grasp fully the premises and the consequences of the new science of mechanics.

Fermat did not publish his analysis of the Menelaean—or Galilean—curve. But Mersenne did, and this, owing to the large circulation of Mersenne's works, is rather important. It seems much more likely that it was Mersenne's drawing of the curve of Fermat—and not that of Kepler—that suggested to Newton his own spiral. The text I have in mind is to be found in Mersenne's *Cogitata Physico-Mathematica* where, in the part dedicated to the study of *Ballistics*, he deals, *inter alia*, with the problems presented by the motions of the projected and falling bodies. The problem of the trajectory of a heavy body falling to the center of the Earth is once more reported,[70] though much more briefly than in the *Harmonie Universelle* and with an important omission: that of his own solution. Instead of it Mersenne adds a report on the spiral line described by the motion of a stone from the circumference of the earth towards its center: [71]

---

[68] *Letter of Galileo to Carcavi* (for Fermat), June 5, 1637. Cf. *Œuvres de Fermat*, Correspondance de Galilée sur la spirale, suppl., 50 sq.: "perche è gran tempo che, havendo con estrema admiriazione veduto et studiato la spirale d'Archimede, la quale ègli compone di due moti equabili, uno retto e l'altro circolare, mi cadda in pensiero la Spirale composta del circolare equabile et del retto accelerato secondo la proportione dell'accelerazione dei gravi naturalemente descendenti, la quale io mi persuado esser in duplicata proporzione di quella del tempo: e questo è la Spirale dell'amico di V.S.

"E sebene nel *Dialogo* vien detto poter esser che mercolato il retto del cadente con l'equabile circolare del moto diurno, si componesse una semicirconferenza che andasse a terminare nel centro della Terra, ciò fa detto per scherzo, come assai manifestamente apparisse, mentre vien chiamato un *capriccio* et una *bizzarria*, cioè *iocularis quaedam audacia*."—As a matter of fact, the spiral of Fermat—something which nobody, until Stefano degli Angeli (cf. *infra*, p. 361) seems to have noticed, is by no means an Archimedean spiral of the second degree; it is, indeed, an inverted spiral.

[69] *Ibid.*, 52: "Aggiungo hora, che sebene dalla composizione del moto equabile orizontale col retto perpendicolarmente descendente con l'accelerazione fatta nella proporzione da me assegnata, si descriverebbe una linea che andando a terminar nel centro sarebbe spirale, niente di meno sin che noi ci trattenghiamo sopra la superfine del globo terrestre, io non mi perito di assegnare a tal composizione una linea parabolica asserendo tali esser la linea che da i proieti vengono descritte. La qual mia asserzione potra somministrar materia d'impugnarmi assai maggiore del moto del mezzo cerchio, il quale almeno facevo pure andare a terminar nel centro, dove anco son sicuro che anderebbero all terminare i proietti. E pur la linea parabolica si va sempre più e più slargando dal asse, che e la perpendicolare al centro." Cf. *supra*, p. 336.

[70] F. Marini Mersenni, *Cogitata Physico-Mathematica, Phenomena Ballistica*, prop. XVIII, pp. 49 sq., Parisiis, 1644.

[71] *Phenomena Ballistica*, prop. XVIII, cor. II, pp. 57 sq. The figure mentioned by Mersenne and given by him on p. 50 is

As Galileo seemed to believe that a stone descending to the center of the Earth (the Earth being supposed moving and replacing the motion of the sun), would move along the circumference B, 3, 4, 5, 6, 7, 8, 9, L about which we have spoken *supra*, the most subtle geometer M. Fermat demonstrated that this descent will not be semicircular but will describe a peculiar spiral, which is the second of the following as the Archimedean one is the first.

that of p. 339 of this paper: *"De linea helice a motu lapidis a terrae circumferentia ad illius centrum descripta.*

"Cum Galilaeus existimare videretur lapidem (posita terra mobili et solis motum supplente) usque ad terrae centrum descendentem, moveri per semicircumferentiam B 3, 4, 5, 6, 7, 8, 9 L de qua superius dictum est, demonstravit acutissimus Geometra D Fermatius non esse descensum illum semicircularem, sed helicem describere peculiarem, quae sit secunda inter sequentes quemadmodum prima est Archimedeae. Sit igitur helix AFB intra circulum BCX descripta, ita ut semper sit eadem ratio circumferentiae BCX ad arcum BC, quae est lineae AB ad FC, vel quadrati AB ad quadratum FC, vel cubi AB ad cubum FC, vel cujuscumque alterius potentiae AB ad similem potestatem FC, regula generalis datur, qua ratio circuli BCX ad spatium linea AB, et helicibus linea AFB comprehensum periatur. Hic apponam octo helices, quarum majores numeri circulorum, minores helicem referunt.

1	3.1
2	15.8
3	14.9
4	43.32
5	33.25
6	91.72
7	60.49
8	153.128

"Quibus placet addere demonstrationem amici, qui demonstravit lineam descensus gravium non esse circularem.

"Sit A terrae centrum, cuius diameter AC; et circumferentia CHB referat Aequatorem: Sint autem spatia, quae per rectam lineam descendentia gravia conficiunt, in ratione duplicata spatiorum quae percurrit in circumferentia: Verbi gratia, si grave primo ex C perveniat ad K, deinde in I, ducunturque rectae AKH, et AIB, ratio HK ad IB erit duplicata rationis arcus CH ad arcum CB. Exempli gratia, si arcus CB duplus est arcus CH, recta IB erit rectae KH quadrupla. Iam vero consideremus cum circulus circa diametrum AC descriptus, per IK puncta transit, recta IB sit HK rectae quadrupla, quando arcus CB duplus est arcus HC. Ducantur recta EK bifariam dividens AC, hoc est ex centro E; et IK, KL ita ut KL sit aequalis AK, et occurat AC producta in L. Cum sint AK, KL lineae a equales, angulus KLC aequalis erit angulo KAL, vel ei aequali IAH; sed anguli $AHIK$, HCL sunt etiam aequales, igitur trianguli AIK, LCK sunt similes.

"Quoniam vero KC latus lateri KI aequalis est, latus CL erit lateri AI aequale.

"Praeterea cum triangula isoscelia AKL, AEK faciant aequales angulos super bases suas aeque sunt similes; et rectangulum LAE erit aequale quadrato. Sed cum reliqua demonstratio pendeat a certis characteribus, quos ille suis usibus accomodabat, qui desunt typographo; sufficiat annotasse lineam istam descensus gravium rectam sub polis futuram, planam helicem sub Aequatore; et in omni alio loco solidam helicem super coni isoscelis superficiem descriptam, cuius basis est parallelus, quo descensus incipit, et vertex ipsum terrae centrum.

"Quam demonstrationem libenter postulantibus communicabo, quemadmodum aliam elegantissimam a D Fermatio inventam, et ad ipsum missam Galilaeum, qua demonstrat spatium ab ista comprehensum helicem esse, vel ad circuli sectorem, vel ad totum circulum, quibus comprehenditur, ut 8 ad 15; quae proportio reperitur similiter inter spatium a spirali circa coni superficiem descriptum, et ipsam coni superficiem."

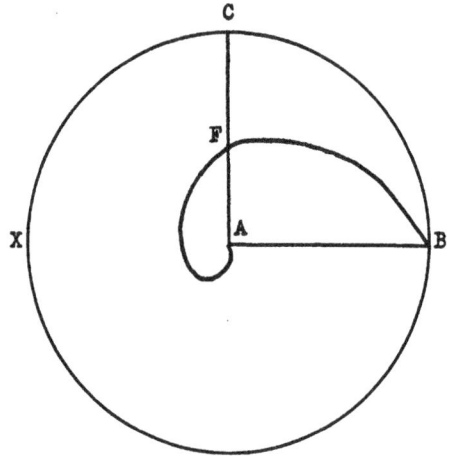

FIG. 9

Let thus the spiral AFB [fig. 9] be described in the interior of the circle BCX, in such a way that the ratio of the circumference BCX to the arc BC be always the same as [that] of the line AB to FC, or of the square AB to the square FC, or of the cube AB to the cube FC, or of whatever power of AB to a similar power of FC; then there will be a general rule, by which the ratio of the circle BCX and the space comprised between the line AB and the spiral AFB will be expressed.

To which it pleases me to add a demonstration of a friend, who demonstrated that the line of descent of the heavy bodies will not be circular.[72]

Let A [fig. 10] be the center of the Earth, of which the diameter is AC; and the circumference CHB represents the equator; let moreover the spaces which the descending heavy bodies traverse on the straight line [towards the center of the earth] be in a duplicate proportion to the distance that they traverse on the circumference: that is, if the body [departing] from C arrives first at K, then at I, and the straight lines AKH and AIB are drawn, the ratio HK to IB will be double the ratio of the arc CH to the arc CB. For instance, if the arc CB is twice the arc CH, the straight line IB will be four times the straight line KH. But let us consider that, if the circle described on the diameter AC passes through the points IK, the straight line IB will be four times the straight line HK, when the arc CB is twice the arc HC. Let us draw the straight line EK, dividing AC in two equal parts, that is, from the center E; and IK, KL such that KL is equal to AK, and touches the prolongation of AC in L. As AK, KL are equal lines, the angle KLC will be equal to the angle KAL, or to [the angle] IAH which is equal to it; but the angles AIK, KCL are likewise equal; therefore the triangles AIK, LCK are similar.

But because the side (latus) KC is equal to the side KI, the side CL will be equal to the side AI.

Moreover, as the isosceles triangles AKL, AEK form equal angles on their bases, they are equally similar and the rectangle LAE will be equal to the square AK.

Yet as the rest of the demonstration depends on certain signs, which he adapted to his usage, and which are lack-

[72] I did not succeed in finding out the name of this friend of Mersenne.

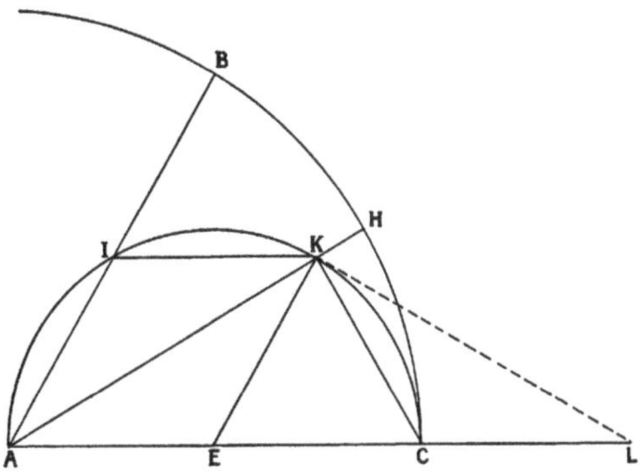

FIG. 10

ing at the printer's, it will be sufficient to mention that this line of descent of the heavy bodies will be a straight line on the poles, a plane spiral on the equator; and in all other places a solid spiral described on the surface of an isosceles cone, whose base is [constituted by] the parallel from which the descent begins, and the vertex, the very center of the earth.[73]

Which demonstration I shall gladly communicate to those who will ask for it, as well as another most elegant one, invented by M. Fermat and sent by him to Galileo himself, demonstrating that the [ratio of] space comprised by this spiral to the [corresponding] sector of the circle, or the whole circle by which it is comprised, is as 8 to 15; which proportion is found to be likewise that of the spiral described on the surface of the cone, and this surface of the cone itself.

IV. BULLIALDUS

Galileo may have held his "semicircular" solution of the problem of the trajectory of the falling body *in hypothesi terrae motae* as being only a "joke." It is, however, quite certain that no one else has ever considered it in this way. Mersenne and Fermat subjected it, as we have seen, to a quite serious criticism. And J. B. Riccioli, though pointing out the incompatibility of Galileo's hypothesis with the indubitable fact, which he had determined experimentally, of the acceleration of falling bodies according to the "duplicate ratio" of time, and also presenting to him some other less important objections, so far accepted it—*in hypothesi terrae motae*—that he based upon it a new objection to the Copernican doctrine of the motion of the Earth.

Yet, before studying the dynamical conceptions of Riccioli which he elaborated in his very influential and widely known *Almagestum Novum* of 1651[74] (and

[73] It is interesting to note: (*a*) that the theory of Mersenne's friend is identical with that of Locher; and (*b*) that Mersenne seems not to have noticed it.

[74] *Almagestum Novum, Astronomiam Veterem Novamque Complectens. . . .* Auctore P. Joanne Baptista Riccioli, S. J. Bononiae, MDCLI. The work was planned to fill three volumes, but only the first one, in two parts, was actually published. However, this small fragment amounts to xlvii + 763 + xviii + 675 pages *in folio*.

reproduced in an abridged and somewhat generalized form in the *Astronomia Reformata* of 1665[75]), we must retrogress a moment to examine the very curious attempt made in 1637 by Bullialdus in his *Philolaus*[76] to develop the Galilean suggestion of excluding the rectilinear motion from nature. By attributing in a hyper-Copernican manner the circular motion as *natural,* not only to the earth but to all bodies that constitute it, he tried to solve the old riddle of the acceleration of bodies during their fall.

Bullialdus starts with an interesting critical exposition of the prevailing theories of his time:[77]

Some people wish the descending body to acquire [in and through its descending motion] a greater gravity [and therefore fall more quickly]. But this is an absurd and insipid explanation which will give no nourishment to the hungry soul. Indeed, gravity, whatever it may be in itself, follows the bulk of matter, and if this remains the same, gravity will remain the same. But from their opinion it would follow that something of [the nature of] a substance can be produced by local motion, which is false, because with the position of local motion is posited only the *ubi* without any new substance resulting therefrom; consequently they assert in vain that it is the gravity acquired by the descent which accelerates the motion.[78]

[75] *Astronomia Reformata,* pt. I, chap. XVII, app. 2, pp. 81–83, Bononiae, 1665. It is interesting to note that a review of the *Astronomia Reformata* was published in the very first volume of the *Philosophical Transactions of the Royal Society,* 394 sq., in 1668. This review, borrowed from the *Journal des Sçavants,* says, *inter alia* (p. 396):

"*Fifthly,* That the Author reasons for the *Immobility of the Earth* after this manner.

"He supposes for certain that the swiftness of the motion of heavy bodies does still *increase* in their descent; to confirm which principle he affirms to have experimented, that if you let fall a ball into one of the scales of a ballance, according to the proportion of the height it falls from, it raiseth different weights in the other scale. For example, a wooden ball of one and a half ounce, falling from a height of 35 inches, raiseth a weight of 5 ounces; from the height of 140 inches, a weight of 20 ounces; from that of 315 inches, one of 45 ounces; and from another of 560 inches, one of 80 ounces, etc.

"From this principle, he concludes the earth to be at rest; for *saith he,* if it should have a diurnal motion upon its center, heavy bodies being carried along with it by its motion, in descending describe a *curve line,* and, as he shews by a *Calculus* made by him, run equal spaces in equal times; whence it follows, that the celerity of their motion would not increase in descending, and that consequently their stroke would not be stronger, after they had fallen through a longer space."

[76] Ismael Bullialdus (Boulliaud), *Philolai, sive Dissertationis de vero Systemate Mundi, libri IV,* Amstelodami, 1639.

[77] I. Bullialdus; *ibid.,* l. I, chap. iv. pp. 11 sq.

[78] Gravity, being an attribute of substance, cannot vary without a corresponding variation of (or in) the substance itself. Such a variation cannot be produced, according to Bullialdus, by local motion, since local motion changes only the place, *ubi,* and not the nature, of the body; *cf. infra,* pp. 348 and 350.

On the scholastic theories of fall and acceleration *cf.,* besides the classic work of P. Duhem, *Etudes sur Léonard de Vinci,* 3 v., Paris, 1909–1913, the very excellent recent publications of Miss Anneliese Maier, *Zwei Grundprobleme der scholastischen Naturphilosophie, Das Problem der intensiven Grösse* and *Die Impetustheorie,* 1939 and 1940; 2nd ed., Roma, 1951 and *Die Vorläufer Galileis im XIV Jahrhundert,* Roma, 1949; *cf.* equally

Others say that heavy bodies, when they tend towards the center, move more quickly and increase their acceleration, because, as the moved [body] traverses more and more parts of the medium, this latter becomes more rare, from which rarefied and extenuated [medium] the moved [body] experiences a lesser resistance; thus, as they say, it is made more rapid in a negative way through the removal of an extrinsic impediment and not through the greater impression of the intrinsic velocity. Moreover they point out that this moving [body] traverses very many parts of the medium, and when it leaves one [of them] other bodies rush in, in order to replenish the space [left by it] which [otherwise] would remain void, when the moving body arrives at the other space. Thus they teach that these bodies [constituting the medium] give to the moving [body] a positive impulse, and being thus impelled, it moves more quickly. Yet it is false that the [medium] is more rarefied by the [moving] body at the end than at the beginning, because [an agent] which remains the same and [acts] in the same [medium] operates always in the same way. But gravity is the same even according to their opinion. Consequently the body descending through the medium rarefies it always in the same degree: and a fluid which yields to a heavy body which presses it because of its gravity, does it neither more nor less at the beginning than at the end of the fall. Further, they commit a *petitio principii* when they assert this proposition, namely that the medium is rarefied by the bodies which traverse it as something admitted and proven; [a proposition] which, moreover, I consider to be false; as neither the fall of a body in water makes it more subtle or rare, nor the fall of a stone from above makes the air more tenuous by agitating it; indeed if the air were rendered more subtle, it would be necessary that the whole sphere of the air likewise became extended, or that some of its parts be extenuated and others be compressed. But who in his senses will say that the stone can either extend the whole sphere of the air or condense some of its parts? Experience makes us certain that water does not extend or occupy more space if, for instance, in a bowl of water is introduced a sphere of lead or of iron: the water will only rise, or, if the bowl were filled up to the extreme brim, there will flow out as much of it as space is needed for the reception of the body; but there will occur no rarefaction and no condensation. The process is exactly the same in the case of the fall of a stone in the air; namely, the air remaining the same in respect of density or rarity yields to gravity and, expelled from one place, occupies the place abandoned by the heavy body.

To the second reason [argument] I answer that the air which moves in, in order to fill the place [abandoned by the body], does not contribute more towards the acceleration of the motion than the air subjected [to the action of the body] resists and opposes it. I add that just as much air is needed to replenish the place in the beginning as in the end [of the motion], wherefore from the beginning to the end [of it] the *impulsus* will be the same [equal]. Besides it is false that the incoming air pushes the descending body, because this air moves into the place relinquished by the body by a contrary movement. The heavy body presses the air which is beneath it and forces it to yield in such a way that it flows to the sides of the body and is borne upwards; by this contrary movement it resists [the downward movement of the body] and by no means accelerates [it]; this selfsame process is manifest also in [the case] of the water; indeed in a [certain] small meas-

ure and proportion it [even] raises up the descending body. Moreover, a manifest contradiction can be seen in the conjunction of these arguments [reasons]: insomuch as the medium is considered as being rarefied it must be considered as less heavy, because [things that are] more rare participate less in gravity; but those which are less heavy likewise press less.

Thus the cause [reason] of the acceleration of the motion of falling [bodies] depends entirely on the natural circular motion, which the part must have in common with the [its] whole, and which beyond doubt the part will keep in its fall, as long as by the violence of its status it is not moved by a violent motion. Accordingly, the circular physical and natural motion moderates the motion of the body, assenting meanwhile to the violent status in which the body is placed, because of which [status] the body, seeking the union with the whole, tends towards it with a certain violence, and therefore is eager to fall upon it by the shortest, i.e., by straight, lines. But that it comes back ever more quickly, that is with an increasing speed, this is produced (caused) by the circular [motion]. And it must not seem strange that these two motions are assigned by nature to one body: this certainly implies no absurdity or contradiction because [both] proceed to the same goal (end). As, indeed, nature does not want to endure violence, or the separation of the parts from the whole, it opposes force by force and violent status by violent motion which, however, is directed and moderated by the natural, in order that nature be restored more quickly. And it is by no means absurd that nature should use violent motion for its restoration. A most excellent example [of such an action] is offered us by the magnet, from which it appears that heavy bodies, which by their nature tend towards the Earth, nevertheless are pulled away [from it] and are raised by the magnet in a violent rectilinear motion, aiming at the reparation of the magnetic nature which wants to join to itself homogeneous [bodies] by an unrestrained *impetus*. This experience must produce and obtain more confidence and authority than a thousand futile fictions of abstract entities. Now, as it is proved by an example that nature acts by violent motion in [things] natural, constituted in a violent status, in order that the natural [one] be restored in the shortest way: it will be our task to prove that the acceleration of this [process of] reparation is performed in no other way than under the direction of the circular motion.

In order to do it we have to prove that a combination of *uniform circular motions*—the only ones that are truly *natural*—can result in a motion that is not only rectilinear but even an accelerated rectilinear one. As a matter of fact, the proof is not difficult to give, particularly since for another purpose it has already been given by Copernicus. Bullialdus, accordingly, in formulating the problem and the theorem of his theory refers quite openly to Copernicus: [79]

To determine two circular motions [performed] in opposite [directions] on two circles, in such a way that the circles be equal and that they have their centers reciprocally on the circumference of each other; and that the motion of one circle, or that of the body moving along it, be double the motion of the center of this same circle on the circumference of the other.

---

my *Etudes Galiléennes* 1, *A l'aube de la science moderne,* and E. Moody, Galileo and Avempace, *Jour. Hist. of Ideas,* 1951.

[79] I. Bullialdus, *op. cit.,* l. I, chap. iv, pp. 13 sq.

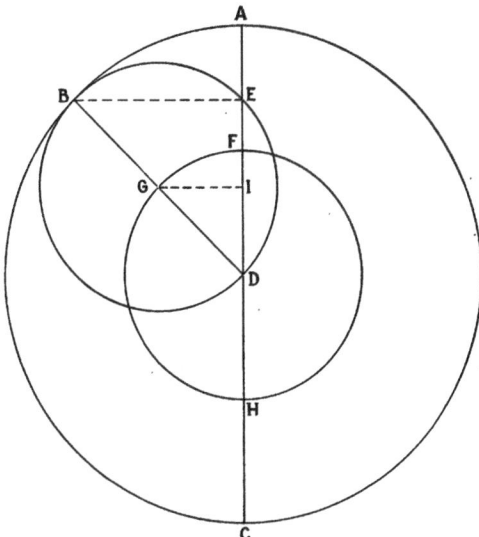

Fig. 11

1. The Theorem

To describe a straight line by means of the above-mentioned circular ones.

Nicolas Copernicus, by whom Astronomy was revived, in the Book on Revolutions III, c. 4, where he deals with the anomalies and computations of the Equinoxes and the obliquity of the ecliptic, and investigates the λοξώσεως [oblique] instability, has demonstrated [that it is] produced by a reciprocal motion composed of circular ones and [that it is] equal (uniform) though nevertheless it appears unequal because of the gemination [combination] of the circular motions. Following him we shall demonstrate it here in our own way.[80]

Let D [fig. 11] be the point from which as a center will be described the circle FGH, the dimension of its semidiameter being DF; then on the periphery FGH let there be taken [arbitrarily] a point, such as G, and this point being taken as center, let there be described the circle BDE, the dimension of its semidiameter being DF or GD; and from the point D let there be described the circle ABC, the dimension of the semidiameter being DGB; finally let there be drawn the diameter ADC which the circle BDE cuts in the point E. But from the point B let [the line] BE be drawn perpendicular to AD, which will be half of the [line] subtending the double [of the] arc BA.[81]

Let it be assumed that at a certain moment of time the line BD will coincide with the line AD, the point B with the point A, and the point G with F; let it be imagined, besides, that while the center G moves from F to G, describing the arc FG of the circle FGH, a certain mobile body will in the same lapse of time move on the circle BED in the opposite direction and with a motion contrary to the motion of G; then it will remain on the line AE and [will] descend on it all the way until C.

[80] Copernicus, in the passage quoted by Bullialdus, demonstrates that a combination of uniform circular motions can produce an alternating rectilinear one.

[81] The line BE will be the sine of BDA.

Now because in the equal circles [GBE and DGF] the arc BE is double the arc GF,[82] the straight line BE [will be] subtending the double of the arc GF and therefore will be double the sine GI. But because the circle ABC is double the circle FGH, the half of the line subtending the double [of the] arc BA will be in a double ratio to the half of [the line] subtending the double arc FG; that is the sine of the arc AB will be double the sine of the arc FG, because these lines are to each other [in the same proportion] as the diameters, that is, as DB is to BE so is DG to GI, or, by substitution, as DB to DG.

But the [line] BE subtends the double [of the] arc FG, therefore it is equal to the sine of the arc AB, or to the half of the [line] subtending the double of BA, which will be true in all the loci. Wherefore the [lines] subtending the arcs of the circle BDE [which are] double the arcs of the circle FG will always fall perpendicularly on the diameter ADC.

But the body moving along BED always traverses circumferences double those [that are described by] the motion of the center G; therefore it will always be found in the terminus of the subtended BE and because of that it will not deviate from the line AC, but will move on it. Which was to be demonstrated.

In the adjoining diagram [fig. 12] let the quadrant $GIKL$ be divided in three equal parts GI, IK, KL, but the quadrant of the exterior circle in the, likewise equal, parts CD, DE, EF, and let the ordinates DH and EG be drawn. Now because the [line] GZ is half of the [line] subtending the double of EF,[83] and the straight line DH the half of the [line] subtending the double [of the] circumference DC, which [DC] is equal to the circumference EF, the straight line DH will be likewise equal to the straight line GZ: because they are the halves of the lines subtending equal arcs which are the doubles of equals. But because the

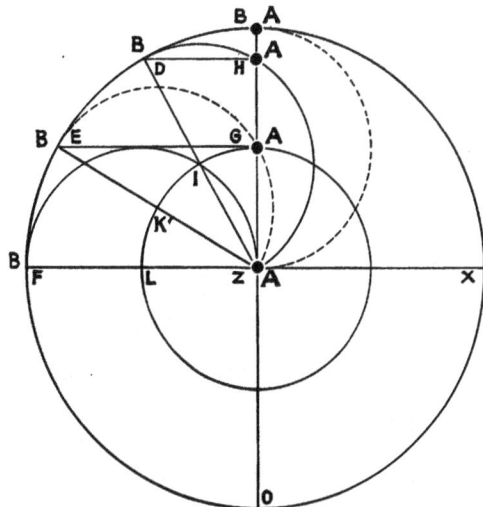

Fig. 12

[82] The arc BE is equal to the arc BA which is twice the arc GF.

[83] The segment EF measures 30°, the double of it, DF, 60°; therefore the line subtending it is equal to the line FZ, which is twice GZ. A point C, missing on the diagram (fig. 12) must be put between A and B.

point H is situated outside of the center of the circle $CDEF$, let us draw through the center Z and the point H the diameter CZO. HZO will be a part of the diameter on which is the center, [and will be] greater than any line drawn toward the periphery from the point H, taken outside of the center; and the remaining part [will be] the least of all the lines drawn [in this way]. Consequently DH will be greater than HC.

But the moving body will describe the line GZ, equal to [the line] DH in the time in which EF or KL will be described by the motion of the center of the circle EGZ; and in the time in which [it] will describe the arc CD equal to the (same) EF, or [in which] the center [will describe] the arc GI equal to KL, the moving body will describe only [the line] CH, which is far shorter than DH or GZ. Thus it is demonstrated that the body A moves more slowly in respect to the circumference than around the center.

The same demonstration will hold concerning the interval HG which, of course, will be greater than the interval CH; indeed to each of the lines CH, HG will be added the equals CZ and twice GZ. The composite line ZCH will be smaller than the composite line HZG. If GZ be subtracted from each, the remainder HZ will be larger than the remainder GCH, because the two [lines] DH, HZ or HZG are greater than DZ or CZ. Consequently the composite [line] DHZ is greater than the composite line ZC, CH; from which DH, ZG and ZC being subtracted, HG will be greater than HC. Thus geometrically the problem and the theorem are demonstrated.

But in order to make an application, let us assume that FZX is the plane of the Earth, as it is conceived in the common way of understanding; and let the stone A be elevated above the Earth by the distance ZC; experience teaches that the velocity [of the fall] will grow [with the descent.] Yet if it descended only on a straight line, and did not perform any other motion, no reason could be given why the descent should be slower at the beginning than towards the end.[84] A mystery of nature worthy of admiration is hidden in this behavior of falling bodies, and it is so pleasant that scarcely anything in the consideration of motion is sweeter. Indeed, the natural motion of the stone in [its] natural state would be on the circumference CDF and would withdraw from the perpendicular as much as it was raised above the Earth; but because of the violent state, Nature submitting to force [re] unites the part with the whole by the straight [line] AZ, and this violence does not prevent the stone [from] moving circularly towards the Earth, and to recede from the perpendicular [path] which is the shortest of all. Indeed, the diameter BZ being uniformly moved with respect to the plane FX, the body A will in some way be moved along the circumference $CDEF$; and thus the body descending by the perpendicular itself will always be found on the semicircle FIZ equal to the quadrant AF, which it will describe in its entirety.

In these two motions the natural one which wants to proceed along CDE, seems to be thrown back in the contrary direction by the violence of the status of the falling body; therefore its revolution is made on a smaller circle, the lines remaining nevertheless equal.

Thus nature has wisely conserved this natural circular motion of the parts of the Earth in their rectilinear descent and their violent status, in order that they may move more quickly to their natural place and are not deprived of their own motion, common with the whole.

The descending body will therefore pass through an infinity of degrees of tardiness and velocity; in the first part of the time [of the descent] it will traverse only the small part CH, in the second, HG; in the third, GZ. Thus to equal parts of time which are represented by the equal arcs CD, DE, EF, correspond unequal parts of the motion, smaller in the beginning and larger towards the end.

You will say that the Earth attracts and the nearer the stone comes [to it], for instance, in falling to the Earth, the stronger is the *impetus* with which it is drawn by the Earth; thus the speed increases without the [above-mentioned] circular motion.

The answer is at hand; I concede that the Earth attracts. But not in such a way that this attraction increases and intensifies the speed. If, indeed, it [the Earth] drew bodies to itself for a magnetic reason, so that they would move less in a greater distance and more in a smaller, then according to this analogy there could exist a distance greater than the semidiameter of the sphere of the activity [of the Earth], in which it could not reunite with itself a part separated from it.[85] But, as we have already said, what exists in truth is not the attraction of the part by the whole, but a combining of all the parts toward their union;[86] as there is a *concursus* of all the lines of the sphere in its centrum, not because of the attractive power of the center, but because of the *consensus*, unique and common to all, of gathering together. How worthy of philosophical considerations are these conjoined accidents of the fall of heavy bodies!

Let me now be permitted to use my mind in a somewhat freer manner and ask the Peripatetic for the freedom (of which to deprive the philosopher is the greatest iniquity) of a closer examination of the convenience of this geometrical proof with reality. May he see that we are not having recourse to figments of things, by which we would explain the not known by the less known. We are not imagining the flight from the void (*fuga vacui*), the danger of which the small body falling through the air will hardly engender and which menace to nature is being averted by the perpetual contiguity of the bodies and the expansibility innate to all fluids, which is such, that the surface of the body moved in a fluid does never, or at most during an indivisible moment, separate itself from the surface of the fluid of air or of water, wherefore the motion of the body in these [fluids] and the contact of the surfaces are simultaneous. The void, indeed, would be most dangerous if between the instant of motion and [that of the] repairing contact [there] could be admitted [to exist] a certain interval of time [but this is not the case].

[Thus] the chimerical child of the new gravity[87] we have bound with chains as a monster that should be thrown into the sea. But the true and genuine cause [of the acceleration] we have illustrated by a geometrical demonstration. Who, if he be not perverted by disgraceful temerity, will deny that the figure, the reason, the mode of motion and its phenomena can and must be explained by geometry?

---

[84] As we have seen, Bullialdus has rejected all the usual explanations of acceleration and has not yet advanced to the principle of inertia.

[85] The action of a finite body, or force, must be finite. Therefore magnetic attraction can extend itself only across a limited distance. A universal magnetic attraction is, consequently, impossible.

[86] It is interesting to note the distinction—even opposition—which exists for Bullialdus (as for Galileo and Borelli) between "attraction of the parts by the whole" which is an external force, and the "tendency towards reunion," which is an internal one.

[87] The new gravity = the gravity that arises from the downward motion of the body.

Bullialdus is obviously very proud of his solution of the old riddle of the acceleration of the heavy bodies in their downward motion toward the center of the Earth. He is even more proud because he obtained this solution by purely geometrical means. Thus he continues:

If it is possible to explain the motions of the skies by reciprocal, conjoined and opposite circular motions, and to demonstrate by geometrical figures their necessary and perpetual restitutions, why should not a similar calculation about the motion of heavy bodies be put in the urn? Well then, if it be permitted to Aristotle to give to the parts of the Earth a straight motion towards the center of the Universe and to deduce therefrom the immobility of the Earth, why should it not be permissible for us who demonstrate that the circular motion is present in the part to deduce therefrom the circular motion of the whole, as well as the motion of heavy bodies, their slowness and their velocity?

One must admire the boldness of Bullialdus who, by adopting the Copernican idea of the natural character of the circular motion—on the Earth as in the skies—and asserting accordingly that the rectilinear one is always and everywhere violent, attempts to turn the tables on Aristotle and to base upon the fall of heavy bodies a proof of the rotation of the Earth. To do so a couple of years after the condemnation of Galileo required, indeed, a rather unusual amount of courage.

Alas, the explanation of the figure and of the mode of motion by geometry is a difficult task and Bullialdus is by no means equal to it. His theory remains as abstract—or even more so—as that of Galileo and Fermat, and no more than they does he attempt to link it to actual data and to examine the problem of the speed possessed or acquired by bodies that fall "to the center of the Earth."

V. RICCIOLI

We must now turn our attention toward Giambattista Riccioli's criticism of both Galileo's and Bullialdus' theories (strangely enough Riccioli mentions neither Mersenne's nor even Fermat's contributions to the problem) in which, quite rightly, he sees an attempt to meet the new and old objections against the movement of the earth, objections already raised by Aristotle and Ptolemy and renewed and modernized by Tycho Brahe. In addition, and he is perfectly right, Riccioli sees in them an attempt to explain, on the basis of the diurnal rotation of the Earth, the famous problem of the acceleration of the downward movement of falling bodies. Giambattista Riccioli, to whom we owe the first careful experimental determination of the value of the acceleration,[88] devotes to this problem a whole section of his *Almagestum Novum*, in which he enquires[89]

Whether, and in which way, by means of the diurnal and also the annual motion of the Earth, an account can be given of the reason for the increment of the velocity of heavy and light bodies (this account being either the only possible, or a better one than all others); so that therefrom this motion [of the Earth] be strongly confirmed. On this occasion the shape [of the line] which, in the hypothesis of the moving Earth, heavy and light bodies describe in their natural motions, is discussed and the argument of Galileo in favor of the motion of the Earth based upon it is definitely destroyed.

I. As the Copernicans have seen [states Riccioli] that, if the Earth be moved in a diurnal and an annual motion, the line which heavy bodies naturally describe in descending and light ones in ascending[90] cannot in reality be a straight line perpendicular to the horizon, such it appears to us, some of the [members] of this sect have worked strenuously in order to determine the shape of this line; and particularly Kepler, though without adding any proof, and Gassendi, deducing it from a similarity with the case of the bullet sent out from a moving motor, thought that this line will be a parabola, or very nearly akin to a parabola, as we have related in chap. 4 of this section.[91] But Galileo and Bullialdus[92] have taught that it will be circular, or composed of circular [ones]: and, not content with that, in the motion accomplished on such a line they believed that they had found the true and solid cause why the motion of heavy bodies appears more and more rapid towards its end, and therefrom they attempted to devise an argument, at first glance most valid, in order to confirm the motion of the Earth. But we have pursued their argument further than they did themselves and have made a hole in the Copernican hypothesis. But we will first deal with what has been invented by Bullialdus.

II. To begin with, Bullialdus, in the fourth chap. of his *Philolaus*, rejects the opinion of those who attempted to assert a cause *a priori* of the greater and greater velocity of heavy bodies towards the end [of their fall], but prin-

---

[88] *Cf.* my paper, An experiment in measurement, *Proc. Amer. Philos. Soc.* **97** (2): 222–237, 1953.

[89] J. B. Riccioli, *Almagestum Novum,* vol. I, pars II, l. IX, sec. iv, chap. iv, and sect. iv, chap. XVII, pp. 398 sq.

[90] Riccioli, in physics, is rather behind the times and still believes—despite Galileo, Stevin, and Benedetti—in the natural lightness of bodies as opposed to their natural gravity. Small wonder that, as we shall see later (*cf. infra*, p. 367, n. 186), he is a partisan of the semi-Aristotelian conception of motion as change produced by an external or internal cause, and that he understands neither the principle of inertia nor that of the relativity of motion.

[91] *Almagestum Novum,* vol. I, pars II, l. IX, sec. iv, chap. iv, pp. 309 sq. (*scholia*). In the *scholia* II and III Riccioli recounts Kepler's theory quoted *supra*, p. 330, n. 4; Kepler, of course, never said that the line of fall will be a parabola. This is only Riccioli's deduction from Kepler's drawing. Gassendi, whom Riccioli discusses in the subsequent *scholia* IV–IX and whose *De motu impresso a motore translato* (Paris, 1642) he cites, never investigated the path of a heavy body descending toward the center of the Earth, but only the trajectory of such a body as it falls from (or is thrown up towards) the summit of a tower or the mast of a moving ship. It is only this portion of the line of descent (or of ascent) that Gassendi, following Galileo, identified with a parabola—an identification which is, of course, practically quite correct. Thus neither Kepler nor Gassendi contributed anything to the problem under discussion. On Gassendi's treatment of the problem of fall, *cf.* my *Etudes Galiléennes* 3: 144 sq. As a matter of fact, it was Gassendi who, *for the first time*, really made the experiment, at least a public one, of throwing a bullet from the mast of a moving ship. Yet it is possible that such an experiment was made by Thomas Digges; *cf.* Francis R. Johnson, *Astronomical thought in Renaissance England,* 164, Baltimore, Johns Hopkins Univ. Press, 1937.

[92] *Cf. supra*, pp. 334 sq. and pp. 347 sq.

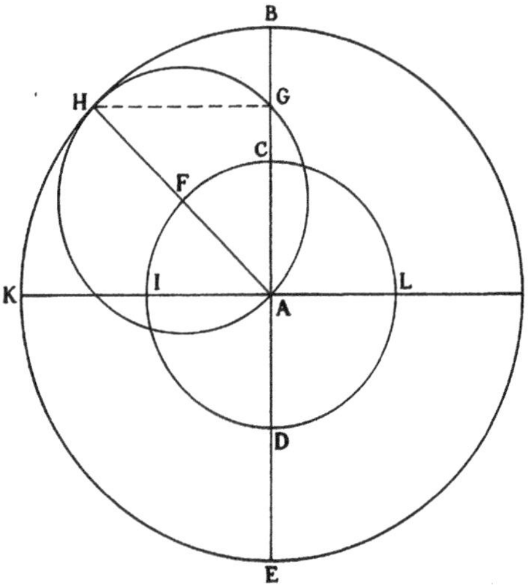

Fig. 13.

cipally [the opinion] of those who have recourse to a greater gravity, or to the second act of gravity and says: "Others desire the descending body to acquire a greater gravity"; which he strongly derides saying: "An absurd and insipid reason laboring under a manifest falsity: gravity, indeed, whatever it may be in itself, follows the bulk of matter; and if this remains the same, the gravity will remain the same. But from their opinion it would follow that something of [the nature of] a substance can be produced by local motion, which is false." Yet, as Scipio Claramontius has very well replied to him in the first part of his *Antiphilolaus*,[93] chap. 4, these authors have not understood that the new gravity acquired by this motion is not in the first act, but [only] in the second act;[94] and therefore many of them call it "gravitation" or "translative impetus downwards." However, having rejected the above-mentioned and other opinions, Bullialdus concludes: "The reason therefore for the acceleration of the motion of falling [bodies] depends entirely on the circular motion which the part must have in common with the whole and which beyond doubt the part will keep in the fall, as long as by the violence of its status it is not moved by a violent motion." He thinks moreover that this motion is accomplished by two opposite circular motions, by which the moving body, in conformity with the appearances, is maintained on the same straight line and appears to move with a non-uniform motion; though in truth each of the circular motions is uniform. In order to prove it he adds the demonstration of Copernicus, chap. 4 of book III, devised for

the [explanation] of the libration and the motion of the Equinoxes, which we have already exposed in book III, chap. 29, n. 10.[95] Yet we must give here a short account of it and apply it to the present question.

III. In the following diagram,[96] describe from the center *A* the circle *BKE* and a smaller one *CIDL*, the semidiameter of which, *AC*, is half the semidiameter *AB*; and on the smaller circle let there be another one, *HGA*, of which the center *F* be deferred in consequence [moved from west to east] by the periphery of the circle *CIDL*, from *C* to *I*. Let, meanwhile, a moving body on the periphery of the circle *AGH* be moved with twice as great a velocity in precedence [from east to west]; the starting points of the motion of the center *F* be *C*, and of the other mobile *H*, be *B*. This being posited, Copernicus shows that by these two opposed motions in contrary directions the moved body will be retained on the straight line *AB* and when the center *F* will describe the semiquadrant *CF*, the body *H*, which has been in *B*, having described the quadrant *HG*, will be in *G*. And when the center *F*, having described the quadrant of its circle *CFI*, is in *I*, it will result that the moving body, having described the semicircle *HGA*, will be in *A*. Hence, if these motions proceed in this way, the mobile will be retained on the straight line *BAE*, and though the motion on it will appear as non-uniform, it will as a matter of fact, be uniform on the above-mentioned circles. Thus the motion of the heavy bodies which to us seems to be performed on a straight line and with a greater and greater velocity is in truth nevertheless composed of two circular and uniform motions.

Sections three and four of the chapter we are dealing with expose the objections made by Scipio Chiaramonti to the theory of Bullialdus. Yet, as they are not only perfectly worthless, but also perfectly traditional and, besides, based on a thorough misunderstanding of the work of the Parisian astronomer; as they are, moreover, rejected as valueless by Riccioli himself, I will not relate them and will proceed with the text of Riccioli, who—rightly—concludes his discussion of Chiaramonti by stating:[97]

V. It is therefore in a different way that Bullialdus must be refuted. And, first of all, if the heavy body would in equal times traverse, in truth, equal distances on its circle, it would strike the Earth with an *impetus* by no means greater if it should fall from a great height than [if it fell] from a low place; which is against most numerous and most evident experiments described in chap. 16 from no. 6.[98] In the second place, from the motion determined by those of the two above-mentioned circles the increment of the velocity manifestly apparent in the natural motion of heavy bodies does not result, which in chap. 16, no. 11 and 12 we have shown, and in no. 24, theorem 3 have proved[99]

---

[93] Scipionis Claramontii S. J. *Antiphilolaus, in quo Philolao redivivo de motu terrae et solis ac fixarum quiete repugnatur,* Cesenae, 1643.

[94] Meaning that the "new gravity" i.e., the "gravity" acquired in and through the very acceleration of downward motion—is not to be considered as belonging to the "first," substantial, actuality of the body, but as something which, though not pure accident, has still only a secondary and derivative reality. This notion is by no means a "scholastic abstraction." It is perfectly analogous to the contemporary distinction between the "proper mass" of a body and its (relativistic) increase as a function of its speed.

[95] *Almagestum Novum*, vol. I pars I, p. 171.

[96] The diagram in question is that of Bullialdus, which I have reproduced on p. 347, but as Riccioli changes the designations, I have to reproduce it anew (fig. 13).

[97] *Almagestum Novum*, loc. cit., 399.

[98] Chapter XVI contains a description of Riccioli's experimental measurement of *g*. On these experiments, cf. my paper quoted *supra*, n. 51.

[99] *Almagestum Novum*, vol. I, pars II, l. IX, sect. iv, chap. xv, n. 24, theorem 3 (p. 394): "Non solum Galilaeus, ut ipse in dialogo 2 de systemate mundi . . . sed et nos certissimis experimentis, per intervalla temporum sensibilia deprehendimus

to follow the ratio of the odd numbers counted from unity. That it [the acceleration] does not follow is easy to demonstrate: indeed, according to Bullialdus, the heavy body B will arrive in A when the circle CFD, proceeding with a uniform motion in consequence [from west to east] together with the circle BKE will describe a quadrant in which time the body [B] on the circle HGA will traverse a semicircle in precedence [from east to west]. Let us divide the whole time of the descent in four equal parts. In the first [period of] time the outermost point H of the diameter AFH will traverse the fourth part of the quadrant, that is 22°30′, which let us imagine to be contained in the arc BH; consequently its versine BG will have 76 of those parts of which the whole sine AB will have 1000; and at the end of the second time it will traverse 45° of which the versine is 293. And at the end of the third time it will traverse 67°30′ of which the versine is 617, and finally at the end of the fourth time it will describe the quadrant BHK, of which the versine AB is the same as the whole sine of 1000 parts. Subtract the first versine from the second, the second from the third, and the third from the fourth, and there will remain the spaces traversed in each of the four times, as one sees in the first column of the following table.[100]

The spaces traversed taken separately	The spaces that should be traversed taken separately	The sequence of odd numbers
76	76	1
217	218	3
324	380	5
383	532	7

But according to the [ratio] of increase of the series of odd numbers which is followed by the velocity of the heavy bodies, if the space of the first time is 76, the spaces of the other times taken separately should be those that you see in the second part of the table. Thus, though the space traversed by means of the circles of Bullialdus in the second [period of] time is nearly equal to the space which should be traversed, however, the third, and much more still, the fourth [spaces] fall behind the due increase in speed. Indeed, the clay ball in our experiments, which I have described in chap. 16, no. 12, traversed in the fourth second of time 240 Roman feet, and in the first second 15,[101] whereas if the ratio of Bullialdus were correct, it would not traverse at the end of the fourth second more than 76 feet. Which is absurd and contrary to most evident experiments. In the third place, by means of the above-mentioned circles Bullialdus cannot give an account of the velocity of light bodies [moving] upwards,[102] as for instance of pure air and of fire, because these are not

praedictam proportionem numerorum pariter imparum ab unitate."
This is all the proof that Riccioli uses to substantiate the theorem that bodies in their natural descending movement follow the law established by Galileo. It is difficult to see the difference made by Riccioli between *ostendere* and *inculcare*.

[100] It is very characteristic of both men, and very interesting to note, that while Bullialdus neither condescends to use actual numbers nor to confront his theory with experimental data, such a confrontation is the first thing that Riccioli has in mind. In the case of the latter, however, one cannot help noticing the utter absence of theoretical discussion.

[101] A determination of 15 Roman feet as the value of the acceleration of—or, more exactly, of the space traversed by—a falling body in the first second of its fall is the best value obtained before Huygens. The roman foot is equal to 29.57 cm.

[102] Riccioli still believed in the qualitative difference between lightness and gravity. Thus he assumed that light bodies accelerate their movement upward in the same ratio as heavy bodies their movement downwards.

moving to the center of their whole, but towards the circumference. Let us therefore hear Galileo, who believes that he has approached nearer to the truth by means of a unique circular motion.

The demonstrations of Galileo by which, the hypothesis of the motion of the Earth being accepted, he attempts to give the reason for the increase of the velocity of heavy bodies.[103]

VI. Galileo, in the second Dialogue on the two Systems of the World starting on page 157 of the Italian [edition] —page 119 of the Latin translation—asserts under the mask of Philip Salviati, the following propositions: *First,* that if the rectilinear motion of heavy bodies toward the center of the Earth were uniform, then, as the motion of the Earth to the east is assumed to be likewise uniform, out of these two motions would be composed a spiral line, namely one of those which Archimedes has defined in the book on spirals; and he said that it is formed by a point moving uniformly on a straight line which is uniformly turning around the other end-point of itself as the center of its rotation. But as the apparent motion of heavy bodies is uniformly difform[104] and continuously accelerated, it is necessary that the line described, [resulting] from this motion [downward] and the diurnal motion of the Earth, recede successively in a greater proportion from the circumference of the circle which the center of gravity would describe if the stone remained on the top of the tower. And, moreover, it is necessary that this withdrawal be at the beginning very small and minimal, because the heavy body passes from rest, that is from the privation of motion downward, and begins the motion downward, and therefore it traverses necessarily all the degrees of slowness interposed between rest and whatever degree of velocity; that these [degrees] are infinite, he has explained beforehand.

Second. As the natural descent of heavy bodies from the summit of the tower to the center of the Earth occurs by itself as towards its goal, he says that it follows necessarily that the line of the composite motion withdraws from the periphery of the circle, described by the summit of the tower, in a greater proportion [than if it moved uniformly] in order that it may terminate in the center of the Earth.

VII. *Third,* that the above-mentioned line is circular, or very nearly approximate to the circular he affirms on page 160 of the Italian edition (p. 121 of the Latin) because this line satisfies the conditions laid down in propositions 1 and 2. Which circular line he teaches on the preceding page to be described in this way:[105] Let A [fig. 5] be the center of the Earth from which, with the distance of the semidiameter BA let there be described the quadrant of the periphery of the Earth, BIM etc. Let there be moreover the tower BC, of which the top C will describe around the center A a quadrant of the circle in the same direction, that is toward the east, to which the Earth moves in its [daily] rotation; let then the whole distance AC be divided in E in two equal parts, and from the center E, with the interval AE, let there be described the semicircle AIC: "I affirm now," says Galileo, "that it can be assumed with a sufficient probability, that the stone, falling from the summit of the tower C, will move by a movement composed

[103] *Op. cit., loc. cit.,* n. 6, p. 399; *cf. supra,* p. 334.

[104] *Uniformiter difformis* = changing in a uniform way, i.e., either uniformly accelerated or uniformly retarded. It is interesting to note Riccioli's use of the old scholastic terminology, current in the Oxford and Paris schools—a terminology completely ignored by Galileo, and, of course by Descartes, but not by Mersenne; *cf. supra,* n. 44.

[105] Riccioli reproduces the figure given by Galileo; *cf. supra,* p. 334, fig. 4, and p. 338, fig. 5.

of the common circular and its own rectilinear." Indeed, if on the circumference *CD* were traced a number of some equal arcs, for example, *CF, FG, GH, HL, LD* and from the center *A* were drawn the straight lines *AF, AG, AH, AL, AD* to the end points of the aforementioned arcs, the parts of these lines intercepted between the two circumferences *CD* and *BI*, that is, *OF, PG, QH, RL, ID*, would always represent to us the selfsame tower, carried around by the rotation of the terrestrial globe toward *D*, [and spaces] by which the stone in its fall will seem to us to descend by a rectilinear motion; which spaces are situated near the base of the tower and are carried on the arc *BI* in the same direction and with a uniform motion. Moreover, the parts of the above-mentioned lines, intercepted between the peripheries *CD* and *CI*, that is *FS, GT, HV, LX, DI* will be the spaces apparently traversed downward by the stone at the end of the single times. And the places in which the stone will be seen at the end of the abovementioned times will be *S, T, V, X, I*, which points recede from the periphery *CD* in an ever-growing proportion; and therefore the motion of the stone will seem to us to be more and more accelerated; and nevertheless we understand that this motion, if nothing prevents it, must end in the center of the Earth *A*, having described the semicircle *CIA*.

VIII. *Fourth*, Galileo affirms, p. 159 (Ital.) and 120 (Lat.), that the natural motion of the stone is, in truth, circular, though it appears to us rectilinear, and that it moves along the simple circumference of only one circle. And that this body will traverse no less space in falling than [it would do] if it remained on the summit of the tower *C*. This because the whole arc *CD*, which it would describe in remaining on the summit of the tower, is exactly equal to the arc *CI* which it describes in falling; just as, taken separately, the arc *CF* is equal to the arc *CS*, and the arc *FG* to the arc *ST*.

Wherefrom follows the third member of this proposition, which he himself calls admirable; namely, that the true and real motion of this stone is not accelerated but is equal and uniform in itself, although it appears to us unequal, because to the equal times in which the summit of the tower is moved, [times] designated by the equal arcs *CF, FG, GH, HL*, correspond equal real spaces *CS, ST, TV, VX, XI*. Which circumstance, says Galileo, liberates us from the labor of investigating the causes of the acceleration of this motion, as the moving body, in remaining on the tower as well as in falling, moves always in the same way, that is, circularly, with the same speed and in a uniform manner. Now, that the above-mentioned arcs are equal, is very easy to demonstrate . . . . [106]

Having accomplished this Galileo concludes, p. 160 (Ital.), p. 121 (Lat.), in these words: "And if concerning the motion of descending heavy bodies we do not say that it occurs precisely in this way, nevertheless I affirm with certainty that if the line described by the falling body is not exactly this same, it is however most near to it." But let us see now the objections—one of them raised by Chiaramonti; the remaining, our own, against Galileo.

The objection of Chiaramonti being, once more, so bad that Riccioli himself is bound to declare it worthless, we shall pass directly to his own:

X. *First*, I will say what I have brought against Bullialdus. If the motion of heavy bodies naturally falling were in truth equal, and appeared unequal solely because

of the deception of the eyes,[107] they would fall from a more elevated place with no greater *impetus* than from a lower one, and therefore the innumerable effects which result from the more vehement percussion of [bodies] falling from a higher place which we have reported in chap. 16 n. 19 would not occur.[108]

Second. By most excellent experiments made before numerous witnesses worthy of confidence (described in chap. 16 n. 13 and n. 14, expanded in various corollaries), it has been shown that if two heavy bodies of different weight are dropped at the same time from the same height, that one which is heavier will descend more quickly, if it is heavier both individually and specifically,[109] as in corol. 4; if it is heavier individually but of equal weight specifically, and if it is heavier specifically but of equal weight as an individual as in corol. 2. But in order that there be no place for any evasion, I will report here the last case, of which I have treated in the same chap. n. 13 exper. 12. I have told there that a wooden ball of 2½ ounces dropped from the crown of the Torre degli Asinelli in Bologna, together with a leaden ball weighing the same 2½ ounces, was 40 feet or 8 paces away from the pavement of the base of this tower when the leaden one was already striking the pavement. They were dropped from the height of 280 feet or 56 paces, and thus the wooden [one] traversed in the same time only 48 paces.[110] From which (probl. 5, example 2 which is to be found in the same chap. 16, no. 29) I have deduced that if such balls were dropped from a height of 929,210 Italian miles,[111] it would happen that at the end of 6 hours the leaden [ball] would strike the Earth, whereas the wooden [one] would be still distant from the Earth by 129,530 miles, that is by nearly 31¼ semidiameters of the Earth.[112] These and

---

[107] It may seem strange that Riccioli appears unable to grasp the meaning of the concept of relative movement in its opposition and contradistinction to that of absolute movement, for the idea of this opposition is as old as mechanics itself (*cf.* P. Duhem, *Le mouvement absolu et le mouvement relatif*, Paris, 1905). Yet for him relative movement is identical with a motion that is merely apparent. It is, therefore, considered as having no reality and thus as incapable of producing a real effect. This blind spot in his work is somewhat surprising since Galileo, whom Riccioli had studied rather carefully, devoted a fair number of pages of his *Dialogue* to an explanation of the difference between the two sorts of movement, and to a demonstration that it is only as relative motion that movement acts and has a physical reality. Riccioli's misunderstanding is, at the same time, revealing: it is a proof of the difficulty of the notion for a seventeenth-century mind.

[108] This conclusion of Riccioli is, of course, completely erroneous, and it is based on his inability to interpret correctly the meaning of Galileo's and Bullialdus's theory; *cf. supra* p. 350 and *infra*, p. 360.

[109] The taking into account of *both* the individual weight and the specific gravity of the body gives a good measure of the clarity and acuteness of Riccioli's experimental thinking.

[110] This "lagging behind" of the wooden (lighter) ball is not considered by Riccioli as an argument against Galileo and in favor of Aristotle. It is explained by him, quite correctly, as an effect of the resistance of the surrounding medium (air) to the motion of the bullets.

[111] An Italian mile = 5,375 feet.

[112] Riccioli assumes (*a*) that the density of the atmosphere is the same throughout the universe, and (*b*) that the acceleration of falling bodies is constant, i.e., is independent of their distance from the earth. The latter assumption is shared by Galileo and Borelli; *cf.* my paper, *La gravité universelle de Kepler à Newton, Archives Internationales d'histoire des sciences*, no. 16, 1951.

---

[106] The demonstration is indeed very easy, so easy that I do not feel I have to reproduce it.

similar experiments [113] are not apt to be explained by the circular motion determined by Galileo, or to be considered as differing from it only slightly (insensibly). Indeed, going back to the figure of n. 7 [114] we see that if both spheres were released at the same time from the summit of the tower C, and each one arrived at the center of the Earth A by the semicircular periphery CIA, it would be necessary that each of the globes, at each moment of the duration of time in which the motion continues, should appear, and even in truth should be, in the same point of the same periphery; thus when the leaden [ball] is in S, the wooden one should likewise be in S, and appear to be there to our eye placed in O. And when the leaden [ball] strikes the surface of the Earth in I, in the same place assuredly the wooden must be. Otherwise, if the wooden [one] were not in the same points of the same periphery, but [in those] of another one, greater, drawn however through the summit C, this circular line would not end in the center of the Earth, but in another point below this center.[115] Which is contrary to the nature of heavy bodies and contrary to the second proposition of Galileo, of which I have spoken in n. 6. But if the wooden globe during the whole time of the motion were in the same points of the same periphery as the leaden one, it would descend with a precisely equal velocity and would not be separated from the Earth by as much as 40 feet. Wherefrom Father Grimaldi has formed a hypothesis, namely that in this same Dialogue 2 about the System of the World Galileo has denied that of two heavy bodies of different weights dropped at the same time from the same point, the one which is heavier will more quickly reach the Earth,[116] just in order that this [difference in speed] should not hinder the motion of heavy bodies on the aforementioned circle invented by him. Yet it is possible that the reason for this denial was [the fact] that he observed two globes of different weight and bulk but of the same kind (species); in this case the difference in the descent and in the percussion appears much smaller than in the other comparisons;[117] and does not manifest itself evidently if [they are] not released from a very great height. But, as it is attested in the same Dialogue, Galileo did not make use of an altitude greater than 100 cubits.[118]

[113] To call the dropping of a wooden ball from the height of 929,210 Italian miles "an experiment" is, perhaps, somewhat pretentious.

[114] Cf. supra, p. 338, figure 5.

[115] This reasoning of Riccioli is perfectly correct: the semi-circular path implies an identical velocity of fall for all bodies, irrespective of their nature; as Mersenne has already seen.

[116] Father Grimaldi is in error; Galileo knew perfectly well that on the earth, in hoc vero aere, heavy bodies fall more quickly than light ones.

[117] The difference in speed (and time) of fall between two lead balls, or between two wooden balls, of different dimensions and weight, is smaller than the difference between that of a lead and that of a wooden ball.

[118] With respect to these Galilean experiments, Mersenne had even doubted that Galileo ever made them at all; cf. Marin Mersenne, Harmonie Universelle, 87, Paris, 1636, where Mersenne expresses his amazement at the very great difference in the figures obtained by him and by Galileo. Cf. also corol. I (p. 112), where, reporting Galileo's description in the Dialogue of the downward motion of heavy bodies on inclined planes, he adds:

"Je doute que le sieur Galilée ayt fait les experiences des cheutes sur le plan puisqu'il n'en parle nullement, et que la proportion qu'il donne contredit souvent l'experience: et desire que plusieurs esprouvent la mesme chose sur des plans differens avec toutes les précautions dont ils pourront s'aviser, afin qu'ils

And one should not say that this inequality could be admitted if the wooden globe were carried around on another periphery slightly larger than the periphery CI; for, if it happened in this way, the two globes would still strike the point at nearly the same moment, nor would the difference of the distances be so sensible as to be 40 feet. Besides if each of the globes were released from the altitude of 929,210 miles, and the medium were of the same density as the air in our vicinity, and if they were not consumed or dissipated by the heat engendered by their motion before their arrival at the Earth, then when the leaden [ball] after the passage of 6 hours reached the center of the Earth, the wooden one would be distant from the center A by $31\frac{1}{4}$ terrestrial semidiameters, as I have shown in chap. 16 n. 29 at the end. Which difference is very remarkable.[119]

XI. Third. The motion of heavy bodies on the circle assigned by Galileo is unable to represent the greatest part of the motions by which heavy bodies descend naturally: not only because he does not explain any other motion but the motion of heavy bodies descending in the plane of the equator,[120] not only because he does not explain the duration of the motion of all [the bodies] but only of those which, because of their levity or the lack of a greater gravity, are so slow that they cannot arrive at the center of the Earth in less than 6 hours' time.[121] But of this kind [of bodies], if the descent should be made through air, there is hardly any to be found; even if it be made through water, heavy bodies which descend so slowly are extremely rare: such is however a sphere of the more rare ebony dropped through water, as it appears from the problem which I have treated in chap. 16 n. 30. But much more numerous are heavy bodies which neither descend on the plane of the Equator nor are so slow that for the traversing of a terrestrial semidiameter, that is of 4,139 miles, they need 6 hours, as it is clear from what is said in chap. 16 n. 25, example 3 and n. 30.[122] I have said [there] first, that the motion of heavy bodies is not explained [by the theory of Galileo] if they do not descend in the plane of the Equator, and therefore if the tower from which the stone falls is not on the very Equator; for the demonstration of Galileo, reported [in] n. 7 and 8 requires that the circle, on the periphery of which CIA the stone is carried around, be in the same plane in which is the center of the Earth A and the summit of the tower C, and therefore likewise the circle CD described by this summit in its diurnal rotation, as it is clear from the construction of the figure and from the hypothesis upon which this demonstration is based; but this can only be the case if this tower is on the Equator: for if it were outside of it, under the poles the descent of the stone would in fact be on a straight line and therefore not on a circle; if, on the other hand, [it were] on some parallel of the Equator, the parallel

voyent si leurs experiences respondront aux notres, et si l'on en pourra tirer assez de lumiere pour faire un Theoreme en faveur de la vitesse de ces cheutes obliques, dont les vitesses pourroient estre mesurees par les differens effets du poids, qui frappera d'autant plus fort que le plan sera moins incliné sur l'horizon, et qu'il approchera davantage de la ligne perpendiculaire."

[119] Indeed!

[120] Cf. supra, pp. 331 sq.: Locher, and pp. 337 sq.: Mersenne.

[121] Mersenne, in his objections to the "semi-circular" theory, admitted implicitly that all bodies fell with the same speed; thus he was enabled to lengthen the time of descent only by increasing the distance to be traversed (cf. supra, p. 341). Riccioli can use as well the variations of speed depending on the weight of different bodies moving in the same resisting medium.

[122] P. 396. By repeating himself, Riccioli makes it unnecessary for me to quote the passage to which he is referring.

described by the foot of the tower would be different from the [one] described by its summit; and besides, the plane of neither would be in the plane in which is the center of the Earth, but in a quite different one. Moreover, Galileo himself in that Dialogue 2, p. 237 (Ital.) and 179 (Latin) when replying to the instances of our Scheiner,[123] recognized that this line would not be circular if [it were] not on the Equator, whereas on the other parallels it would describe a conical surface; for, stimulated by this interrogation of Scheiner which we find in the *Disquisitiones* p. 31 and which is as follows: "Quare centrum spherae delapsae sub Aequatore spiram describit in eius plano? Sub alijs parallelis spiram describit in cono? Sub polo descendit in axe lineam giralem decurrens, in superficie cylindrica consignatam?"—he answers in the following words: "Because of the lines drawn from the center to the circumference of the sphere, which are those by which *graves* descend, that which terminates in the Aequinoctial designeth a circle, and those that terminate in other parallels describe conical superficies; now the axis describes nothing at all, but continueth in its own being."

Wrongly, therefore, did he, p. 121 (Lat.) and 160 (Ital.), assert in a universal and indefinite manner that this line will be circular, and did not correct his assertion, having as I believe forgotten what he had said. I have said *secondly* that not even on the Equator would the motion of heavy bodies be saved, provided that they are not so slow that for their arrival at the center of the Earth (if this arrival be made through a tube full of air or of water) they need precisely 6 hours. For, in order that the demonstration of Galileo which I reported in n. 7 and 8 be correct, it is necessary that the circumference of the circle *CIA* upon which the heavy body is carried around be equal to the quadrant *CD*, etc., which the summit of the tower describes because of the diurnal rotation; and that when the heavy body arrives in *A* by the arc *CIA* the summit of the tower meanwhile accomplishes a quadrant of the diurnal rotation, that is, in the space of 6 hours. Otherwise, if the heavy body arrived in *A* by the arc *CIA* more quickly or more slowly than in six hours, the whole demonstration of Galileo would tumble down. Yet because in reality these two conditions can be realized, namely that the place from which the heavy body is released be on the equator and that [the body] be of such a slowness that it needs 6 hours in order to traverse the semidiameter of the Earth, we shall raise here the strongest argument against the diurnal and also against the annual motion which cannot exist without the diurnal. And this is that which follows. . . .

The strongest argument of Riccioli turns out to be the same as he had already used against Bullialdus and Galileo,[124] namely that if the Earth were rotating and the bodies, accordingly, moved with a uniform speed along a circular line—as Galileo taught they will do—then they would strike the Earth with the same velocity, which means with the same force or momentum. And as this is not the case, it follows therefrom that the Earth does not move.[125]

The argument of Riccioli appears to be rather good. Why, indeed, should a body moving *with a constant speed* increase its momentum with the distance traversed in its motion? Is it not evident that, quite on the contrary, it should conserve it and strike the Earth, or any other goal, always with the same force, irrespective of the place where it does it? A conclusion which would be perfectly irrefutable if the descending body described, with a uniform speed, Galileo's semicircular path on an Earth standing still; in this case, undoubtedly, it would strike the Earth—far away from the tower—with the same tremendous power at the beginning as at the end of its course.[126] If it behaves differently it is just because the Earth does not stay still but moves. Riccioli seems to forget that the Earth is moving along. Or, still worse, he seems to be unable to understand that a uniform circular motion may, combined with some other, not only *appear* but *be* an accelerated one, as he is unable to understand (in the case of Bullialdus) that two uniform circular motions may combine into a rectilinear and accelerated one. This inability, however, is not to be taken as demonstrating Riccioli's obtuseness or stupidity, but only the difficulty of grasping new ideas, or even of adapting old ones—such as the relativity of motion—to new conceptions, such as the motion of the Earth.[127]

The argument of Riccioli, when first published, did not raise a discussion. But when he reproduced it in 1665 in his *Astronomia Reformata* [127a] and even somewhat earlier as Stefano degli Angeli and Riccioli himself will tell us,[127b] it met with a rather strong opposition. A polemic, at first restricted to letters and oral discussion, then public, took place, of which James Gregory published a long account in the first volume of the *Philosophical Transactions* of the Royal Society.[128]

As this account, in spite of having been quoted by Rigaud,[129] seems to have escaped the notice of Newton's historians—probably because the *Philosophical Transactions* is rather difficult to get—it seems to me worth while to reprint it here in full.

VI. GREGORY'S REPORT

An Account *of a Controversy* betwixt Stephano de Angelis, *Professor of the Mathematics in* Padua, *and* Joh. Baptista Riccioli *Jesuite; as it was communicated out of*

---

[123] *Almagestum Novum, loc. cit.,* 400; cf. *supra,* p. 331. As I have noted, Riccioli—though giving the page of the *Disquisitiones*—quotes Locher, calling him Scheiner, not from the *Disquisitions* themselves, but from Galileo; cf. *supra,* n. 11 where I am quoting the text of Locher himself.

[124] Cf. *supra,* pp. 351 sq.

[125] *Almagestum Novum, loc. cit.,* 401.

[126] It would move, indeed, from start to finish with the speed of the earth's rotation.

[127] It is interesting to note that Riccioli's arguments impressed even so acute a mind as James Gregory! Cf. *infra,* p. 356.

[127a] In his *Apologia* (cf. *infra,* p. 391) Riccioli will tell us that in the *Astronomia Reformata* he extended his argument to curves other than the circle; as a matter of fact, he had done it already in the *Almagestum Novum*; thus the *Astronomia Reformata* brings nothing new into the discussion.

[127b] Cf. *infra,* pp. 380 sq. and pp. 392 sq.

[128] The Royal Society, *Phil. Trans.* 1: 684 sq., 1668.

[129] S. P. Rigaud, *Historical Essay on the First Publication of Sir Isaac Newton's Principia,* Oxford, 1838.

their lately Printed Books, by that Learned Mathematician Mr. Jacob Gregory, *a Fellow of the* R. *Society.*[130]

[130] The Royal Society, *Phil. Trans.* 1: 693 sq. The books discussed in Gregory's review—he does not cite them by title—are:

(1) Stefano degli Angeli, *Consiglierationi / sopra la forza / di alcune raggioni / fisicomattematiche / addotte dal M.R.P. / Gio. Battista Riccioli / Della Compagnia di Giesù nel suo Almagesto Nuovo / et Astronomia Riformata contro il / Sistema Copernicano / espresse in due dialogi da F. / Stefano De Gli Angeli / Venetiano, /Mattematico nello Studio di Padoua.* (Appresso Bartolo Bruni, Venetia, 1667.)

(2) Michele Manfredi, replying to Angeli in the name of Riccioli, who did not want to enter himself in the polemics, at least not under his own name [according to Sommervogel, *cf.* Carlos Sommervogel S.J., *Bibliothèque de la Compagnie de Jesus*, s.v. "Riccioli," 6: 1803, Bruxelles—Paris, 1895, "Manfredi" is only a pseudonym of Riccioli], *Argomento / fisicomattematico / del padre / Gio. Battista Riccioli / Della Compagnia di Giesù / contro il moto diurno della terra / Confermato di nuouo con l'occasione della Risposta alle Considerazioni sopra la Forza del detto Argomento, etc. / Fatte dal M. R. Fr. Stefano De Gli Angeli / Mattematico nello Studio di Padoua / All' Illustriss. Signore Il Sig. / Co: Francesco Carlo / Caprara / Conte di Pantano, Gonfaloniere di Giustizia del / Popolo, e Comune di Bologna.* (Per Emilio Maria, e Fratelli de' Manolesi, in Bologna, 1668.)

(3) Stefano degli Angeli, defending himself against Manfredi, and counterattacking: *Seconde / considerationi / sopra la forza / dell'argomento fisicomattematico / Del M. Reu. P. / Gio. Battista Riccioli / della Compagnia di Giesù, / contro il moto diurno della terra / Spiegato dal Sig. Michiel Manfredi nelle sue Risposte, e / Riflessioni sopra le prime Considerationi / di / F. Stefano De Gl'Angeli / Veneriano / Mattematico nello Studio di Padoua / Espresse da questi in due altri Dialogi III, e IV.* (Per Mattio Bolzetta de Cadorini, in Padoua, 1668.)

Besides the books reported on by Gregory there are five others on the same subject:

(4) *Risposta / di Gio. Alfonso / Borelli / Messinese Matematico dello Studio di Pisa / Alle Considerazioni fatte sopra alcuni luoghi del suo / Libro della Forza della Percossa Del R.P.F. Stefano De Gl'Angeli / Matematico nello Studio di Padoua, / All' Illustrissimo, e Dottissimo Sig. / Michel Angelo Ricci* (Messina, 29 Febraro, 1688).

(5) Stefano degli Angeli, *Terze / Considerationi / Sopra una lettera / Del Molto Illustre, et Eccellentissimo Signor / Gio. Alfonso Borelli / Messinese Mattematico nello Studio di Pisa, / Scritta da Questi in replica / Di alcune dottrine incidentemente tocche / Da Fra / Stefano De Gl'Angeli / Venetiano / Mattematico nello Studio di Padoua / Nelle sue prime Considerationi sopra la forza di certo Argomento / contro il moto diurno della Terra, / Espresse da questo in un Dialoge / Quinto in ordine.* In Venetia M.DC.LXVIII / (Appresso li Heredi Leni con licenza de' Superiori).

The *Terze Considerationi* are not listed by Sommervogel.

(6) Borelli's defense by Diego Zerilli: *Confermazione / d'una sentenza / del Signor / Gio. Alfonso / Borelli M. / Matematico dello Studio di Pisa / di nuouo contradetta / Dal / M.R.P. Fra Stefano / De Gl'Angeli / Matematico dello Studio di Padoua / nelle sue terze considerationi / prodotta da / Diego Zerilli.* (In Napoli, per Ludovico Cauallo, 1668.)

(7) *Apologia / R.P. / Io. Bapt. Riccioli / Societatis IESU / Pro Argumento / Physicomathematico / Contra Systema Copernicanum / Adiecto contra illud nouo Argumento ex Reflexo / motu Grauium Decidentium / . . .* (Venetiis, MDCLXIX. / Apud Franciscum Salerni, et Joannem Cagnolini / Superiorum permissu.) Sommervogel quotes the *Apologia del R. P. Gio. Battista Riccioli*, but curiously enough, pretends it to be the same work as that of Manfredi.

Riccioli in his *Almagestum Novum* pretends to have found out several new demonstrative Arguments against the Motion of the Earth. *Steph. de Angelis*, conceiving his Arguments to be none of the strongest, taketh occasion to let the world see, that they are no more esteem'd in *Italy*, than in other places. *Manfredi*, in behalf of *Riccioli*, endeavours to answer the Objections of *Angeli*, and this latter replyes to *Manfredi's* Answer. The substance of their discourse is this following.

Although the Arguments of *Riccioli* may be many, yet the strength of them consists chiefly in these three:

The *First*.

Multa corpora gravia, dimissa per Aerem, in Plano Aequatorio existentem, descenderent ad Terram cum Velocitatis Incremento reali et notabili; et non tantum apparenti. Sed si Tellus moveretur motu diurno tantum circa suum centrum, nulla corpora gravia, dimissa per Aerem, in Plano Aequatoris existentem, descenderent ad Terram cum velocitatis incremento reali ac notabili, sed tantum cum apparenti. E [rgo] Tellus aut non movetur, aut non movetur diurno tantum motu.[181]

The *Second*.

Si Tellus moveretur motu diurno, aut etiam annuo, multo debilior esset ictus Globi bombardici explosi in Septentrionem aut Meridiem quam ab Occidente in Orientem. At consequens est falsum. E [rgo] et antecedens.[182]

The *Third*.

Si Tellus diurna revolutione moveretur, Globus argillaceus unciarum 8 ex altitudine Romanorum pedum 240. per aerem quietum dimissus, obliquo descensu in Terram delabetur absque incremento reali ac physico velocitatis,

(8) Stefano degli Angeli, *Quarte / Considerationi / Sopra la confermatione / D'una sentenza del Sig. Gio Alfonso Borelli M. / Matematico nello Studio di Pisa / Prodotta da Diego Zerilli / contro le terze Considerationi / Di Stefano degli Angeli etc. E sopra l'Apologia del M.R.P. / Gio. Battista Riccioli / Della Compagnia di Giesu' / A favore d'un suo Argomento detto Fisico-Matematico / Contro il sistema Copernicano / Espresse dal medemo Stefano de gl'Angeli Venetiano Matematico / nello Studio di Padoua in due Dialoghi VI. e VII.* (In Padoua, Per Mattio Cadorin detto Bolzetta, 1669, con Licenza de' Superiori.) Not listed by Sommervogel.

[181] Since Gregory's text is a mixture of Latin and English, I have printed it as it was published by him, giving the translation in footnotes (p. 694): "Heavy bodies let fall through the air in the plane of the equator descend towards the Earth with an increased speed that is real and notable, but only not apparent. But, if the Earth did move in a diurnal motion only around its center, no heavy body, let fall through the air in the plane of the equator, would descend towards the Earth with a real and notable increase of speed, but only with an apparent one. *Ergo*, the Earth either does not move, or does not move in a diurnal motion only."

The last part of the sentence, which Gregory does not explain, is probably a hint at Roberval's attempt to explain the acceleration in the downward movement of heavy bodies by a combination of the diurnal and the annual motion of the Earth. Since nobody but Riccioli (not even Mersenne) ever took this theory into account, it did not seem necessary to study it here.

[182] *Ibid.*: "If the earth moved in a diurnal, or also in an annual, motion, the impact of a cannon ball shot towards the North or the South would be much weaker than that [of the ball shot] from the West to the East. But the consequence is false. Therefore the premise is false too." Riccioli's argument is a modified (and weakened) form of the Tychonian one. *Cf.* my *Etudes Galiléennes* 3: 22 sq.

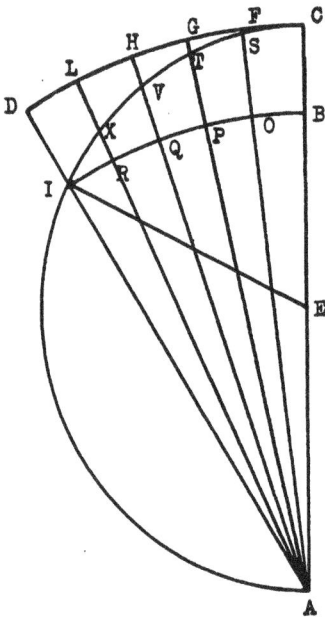

Fig. 14.

vel certe numquam tanto, quanta est proportio percussionis ac soni per casum ex dicta altitudine facti. Sed posterius est absurdum. E [rgo] et prius.[133]

In Answer to the first of these Arguments, *Angeli* denies the *minor*, which *Riccioli* pretends to prove thus:

Si Tellus moveretur solo diurno motu, aliquod grave, dimissum ex Turris vertice *C* in plano Aequatoris existentis, describeret suo motu naturali portionem lineae *CTI*, quae esset ad omnem sensum circularis.[134] [Fig. 14.]

This Angeli denies showing by Computation, that Riccioli his Observation proves no such thing. For (saith Angeli) according to Riccioli, in one second of an hour the weight descends 15 foot; in 2 seconds, 60 foot; in 3 seconds 135 foot and so continually the spaces from the beginning are in duplicate proportion of the Time from the beginning; and according to the same Author, *AB* (the *semidiameter* of the Earth) is of 258700000 foot, and *BC* (the height of the Tower of the *Asinelli* in *Bononia*) of

133 *Ibid.*: "If the earth moved in a diurnal revolution, then a ball [weighing] 8 ounces, released at an altitude of 240 Roman feet, would fall towards the earth in an oblique descent, without any real and notable increase in speed, or [at least] certainly not with [an increase of velocity corresponding] to the proportion of the [force of] percussion and sound, achieved through the fall from the abovementioned altitude. But the conclusion is absurd. *Ergo,* its premise."

Riccioli's third argument is only the first one in another form. But, as we have seen, the learned Jesuit wanted to make the most of his experiments on the fall of bodies in which he established the conformity of the Galilean law (duplicate proportion of speed to time) with the observable facts.

134 *Ibid.,* 695: "If the earth moved in a diurnal motion only, a heavy body, let fall from the summit of a tower *C* in the plane of the equator, would describe by its natural motion a part of the line *CTI*, which is, practically, circular."

240 foot; and therefore *AC* is as 258700240, which hath the same proportion to *FS*, 15 foot, to wit, ye fall in one second, which *AC* in parts 20000000000 hath to *FS* 11596 $\frac{54356}{224189}$; but supposing, with Riccioli, *CSIA* a semicircle, *FS* is 53 parts of which *AC* is 10000000000: Hence concludeth *Angeli*, that *CSIA* is no wayes near to a *semicircle;* which is most sure, if so be the weight falls not to the Center of the Earth precisely in 6 hours: For, in this case of *Riccioli*, the weight falls to the Center of the Earth in 21 *minutes* and 53 *seconds.*[135]

Manfredi in his Answer for *Riccioli* affirms, that *Angeli* understands not the Rule of Three, in giving out *FS* for 11596 $\frac{54356}{224189}$, of which *AC* is 20000000000: And *Angeli* in his Reply affirms his Analogy to be so clear, that there can be nothing said more evident than it self to confirm it; referring in the mean time to the further determination to geometers.

Angeli might have answer'd *Riccioli's* Argument, granting the weight to move equally in a *semicircle,* by distinguishing his *Minor* thus:

Nulla corpora gravia descenderent ad Terram cum velocitatis incremento reali ac notabili, si Velocitas computetur in circumferentia semi-circuli; *Minor* propositio est vera. At non computatur ita motus descensivus: nam hic motus aequalis in circumferentia semi-circuli *CIA,* componitur ex motu aequali in quadrante *CD,* et motu accelerato in semidiametro mobili *CA,* et hic motus acceleratus in semi-diametro est verus et simplex motus descensivus; in qua acceptione Minor proposito est *falsissima* et Riccioli *etiam experientiis contraria.*[136] But it seems that Angeli answers otherwise, to make *Riccioli* sensible, that *CIA* is no semi-circle; concerning the nature of which Line they debate very much throughout the whole discourse.[137]

The *second* Argument is much insisted upon by *Angeli,* to make his solution clear to vulgar capacities; but the substance of all is, that the Cannon-ball hath not only that violent motion, impressed by the Fire, but also all these motions proper to the Earth, which were communicated to it by the impulse received from the Earth: for the Ball, going from West to East, hath indeed two impulses, one from the Earth, and another from the Fire; but this im-

135 *Ibid.,* 695–696. Angeli's arguments are practically the same as those of Mersenne (*cf. supra,* p. 337) whom, by the way, he does not quote.

136 *Ibid.,* 696: "*No heavy bodies would descend toward the earth with a real and notable increase in velocity* if the velocity were to be measured along the circumference of the circle. The *minor* proposition is true. But the motion of the descent is not measured in this way: indeed, this uniform motion on the circumference of the circle *CIA* is compounded by the uniform motion on the quadrant *CD* and by the accelerated motion along the mobile semidiameter *CA,* and this accelerated motion along the semidiameter is a true and simple descending motion. This understood, the minor proposition is most false and *even contrary* to Riccioli's *experiences.*" Gregory, who understands clearly the meaning of the relativity of motion, is, nevertheless, obviously making the same error as Fermat and Angeli. He interprets the downward motion of the heavy body as a motion along a moving radius, i.e., kinematically and not dynamically. It is not impossible that Gregory was acquainted with Fermat's solution of the problem. This is even very probable, although he should then be expected to mention that the line of descent will not be a circle, but a spiral, as Angeli describes it.

137 This debate is very interesting, and I shall report on it *infra.*

pulse from the Earth is also common to the mark, and therefore the Ball hits the mark only with that simple impulse received from the Fire, as it doth being shot towards the North or South; as *Angeli* doeth excellently illustrate by familiar examples of Motion.[188]

To *Riccioli* the *third* Argument Angeli answereth, desiring him to prove the sequel of his *Major*, which *Riccioli* doeth, supposing the *curve* in which the heavy body descends, to be composed of many small right lines; and proving, that the motion is almost always *equall* in these lines; and after some debate, concerning the equality of motion in these right lines, *Angeli* answers, that the *equality* of *motion* is not sufficient to prove the *equality* of *percussion* and sound, but that there is necessary also *equal angles* of *incidence;* which in this case proveth to be very *unequal*. To illustrate this more, let us prove, that, other things being alike, the proportion of two percussions is composed of the direct proportion of their *velocities*, and of the direct proportion of the *Sines* of their angles of incidence.[139]　*Supponamus* autem sequens principium, nempe

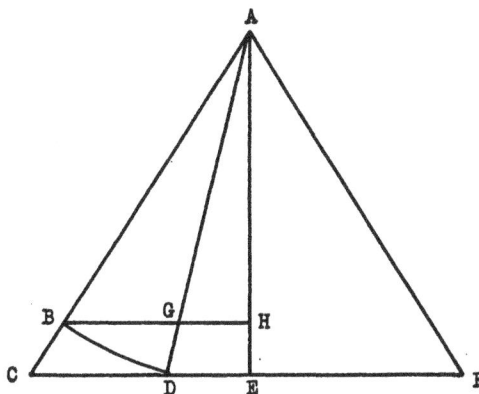

FIG. 15

---

[188] Stefano degli Angeli, as is only natural (and correct), tries, though unsuccessfully, to explain to Riccioli the meaning of the relativity of motion *in hypothesi terrae motae*.

[139] *Ibid.*, 697–698: "Let us assume the following principle: namely, that percussions (*caeteris paribus*) are in a direct proportion to the velocities with which a moving body approaches a resisting plane (fig. 15). Let the plane be *CF*, and let two moving bodies, equal and similar in all respects, move from the point *A* to the plane *CF* in a uniform motion and by the straight lines *AD* and *AF*. I say that the [force of the] percussion in the point *D* to [that of] the percussion in the point *F* will be in a ratio composed from the ratio of the velocity along the straight line *AD* to the velocity along *AF*, and from the ratio of the sine of the angle *ADE* to the sine of the angle *AFE*. Let the straight line *AE*, from the point *A* to the plane *CF*, be a normal one, the straight line *AC* equal to the straight line *AF*, and *AB* equal to the straight line *AD*, and the plane *BGH* parallel to the plane *CF*; let us assume that a moving body, equal and similar to the former ones, moves with a uniform motion along the straight line *AC*, with the same speed with which the body moves on the straight line *AD*: since the planes *BGH* [and] *CF* are parallel, and the motion on the straight line *AC* is uniform, it follows that the moving body arrives at the plane *BH* with the same velocity as at the plane *CF*, and therefore that the [force of the] percussions in the points *B* [and] *C* are equal; and the [force of the] percussion in the point *D* is to [that of] the percussion in the point *B* as the straight line *AE* is to the straight line *AH*, or (because of the equality of the straight lines *AD* and *AB*) as the sine of the angle *ADE* is to the sine of the angle *ABH*; this I prove as follows: the speed of the body on the straight line *AD* is equal to the speed of the body on the straight line *AB*, which is equal to *AD* itself, and therefore both the straight lines *AD* and *AB* will be traversed in the same time; thus the accessions to the resisting planes *CF* and *BH* will be accomplished in the same time; and therefore the velocities of the approaches to the resisting planes are in the direct ratio of *AE* to *AH*, and because of this fact the [force of the] percussion in the point *A* is to [that of] the percussion in the point *C* in the same ratio of *AE* to *AH*; that is, [in the ratio] of the sine of the angle of incidence *ADE* to the sine of the angle of incidence *ACE*, or *AFE*. But because the straight lines *AC* and *AF* are inclined at equal angles to the plane *CF*, bodies moving on the straight lines *AC*, *AF* approach the plane *CF* in the same proportion in which they move on the straight lines *AC* and *AF*. And therefore the [force of the] percussion in *C* is to [that of] the percussion in *F* in the ratio of the velocities of the motion on *AC* or on *AD* to the velocity of the motion on *AF*. But it has already been demonstrated that the [force of] the percussion in

quod percussiones, (caeteris paribus) sint in directa proportione cum velocitatibus, quibus mobile appropinquat planum resistens. [Fig. 15.] Sit planum *CF*, sintque duo mobilia omni modo aequalia, et similia, quae motu aequali accedant a puncto *A*, ad planum *CF*, in rectis *AD*, *AF*; dico, percussionem in puncto *D* ad percussionem in puncto *F* esse in ratione composita ex ratione velocitatis in recta *AD*, ad velocitatem in *AF*, et ex ratione sinus anguli *ADE* ad sinum anguli *AFE*. Ex puncto *A* in planum *CF* sit recta *AE* normalis, sitque recta *AC* aequalis rectae *AF* et *AB* aequalis rectae *AD* et planum *BGH*, parallelum plano *CF*: supponamus mobile, prioribus simile et aequale, moveri aequaliter in recta *AC*, eadem velocitate, qua movetur mobile in recta *AD*: quoniam plana *BGH*, *CF*, sunt parallela, et motus in recta *AC* est aequalis, igitur mobile eadem velocitate accedit ad planum *BH*, qua ad planum *CF*, et proinde percussiones in punctis *B*, *C*, sunt aequales; atque percussio in puncto *D*, est ad percussionem in puncto *B*, ut recta *AE* ad rectam *AH*, seu (ob aequales rectas *AD*, *AB*) ut sinus anguli *ADE* ad sinum anguli *ABH*, quod sic probo; velocitas mobilis in recta *AD*, est aequalis velocitati mobilis in recta *AB*, ipsi *AD* aequali, et ideo eodem tempore perficitur utraque recta *AD*, *AB*; et proinde eodem tempore perficiuntur accessiones ad plana resistentia *CF*, *BH*; ideoque velocitates accessionum ad

the point *C* is to the percussion in the point *D* as is the sine of the angle *ADE* to the sine of the angle *AFE*, and now it is demonstrated that the percussion in the point *C* is to the percussion in the point *F* as is the speed of the motion on *AD* to the speed of the motion on *AF*. Accordingly, from definition 5 of the sixth book of the Elements, [it follows] that the [force of] the percussion in *D* is to [that of] the percussion in *F* in the ratio composed from the ratio of the sine of the angle of incidence *ADE* to the angle of incidence *AFE*, and the ratio of the velocity on *AD* to the velocity [on] *AF*; which had to be demonstrated.

"Let nobody be affected by the fact that this demonstration is restricted to the [case of] uniform motions on straight lines and resisting planes, because it is true in any case. Indeed, as percussions occur in the point, in this the straight, the curved, the uniform and the non-uniform coincide; but if the percussions do not occur in points, about these there cannot be geometrical conclusions. [In these cases] the defect of the conclusion is to be evaluated according to the defect of the matter from the required conditions, as it must always be done when geometrical demonstrations are applied to a physical body."

plana resistentia sunt in directa ratione AE ad AH, atqui ideo percussio in puncto D est ad percussionem in puncto C in eadem ratione AE ad AH; nempe ut sinus anguli incidentiae ADE, ad sinum anguli incidentiae ACE, vel AFE. Quoniam autem rectae AC, AF aequaliter, inclinant ad planum CF, mobilia in rectis AC, AF accedunt ad planum CF, in eadem proportione qua moventur in rectis AC, AF; et ideo percussio in C est ad percussionem in F in ratione velocitatis motus in AC seu in AD ad velocitatem motus in AF; At demonstratum est ante, percussionem in puncto C, ad percussionem in puncto F esse in ratione sinus anguli ADE ad sinum anguli AFE, et nunc demonstratum est, percussionem in puncto C esse ad percussionem in puncto F, ut velocitas motus in AD ad velocitatem motus in AF. Igitur ex 5 defin. 6 Elementorum, percussio in D est ad percussionem in F, in ratione composita ex ratione sinus anguli incidentiae ADE, ad sinum anguli incidentia AFE, et ex ratione velocitatis in AD ad velocitatem in AF; quod demonstrare oportuit. Neminem moveat, quod haec demonstratio adstricta sit motibus aequalibus in lineis rectis et planis resistentibus; est enim vera in omni casu: nam, cum percussiones fiant in puncto, in hoc coincidunt rectum, curvum, aequale, et inaequale; si autem in punctis percussiones non fiant, de illis non potest dari consideratio geometrica, sed judicandus est conclusionis defectus secundum defectum materiae a conditionibus requisitis, sicut semper fieri debet, dum demonstrationes geometricae corpi physico applicantur.[140]

In *Angeli* his reply to *Manfreddi*, he maketh mention of an Experiment, which (as was related to him by a *Swedish* Gentleman) had been made with all due circumspection by *Cartesius* to prove the *Motion* of the *Earth*. The experiment was; He caused to be erected a Canon perpendicular to the Horizon; which being 24 times discharged in that posture, the Ball did fall 22 times towards the *West*, and only twice toward the *East*.[141]

VII. BORELLI

I do not need to insist upon the interest and historical importance of James Gregory's account of the polemic around R. P. Riccioli's attempt to use the Galilean law of fall as a basis for refutation of the Copernican astronomy. I have to point out, however, that Gregory's communication to the Royal Society [142] indicates quite clearly that there was much more in this Italian discussion than he reports on, namely, a discussion about the trajectory described by a body falling from a high tower *in hypothesi terrae motae*—a problem in which, unfortunately, James Gregory, unlike ourselves, happened not to be interested.

Thus it appears that we cannot rely solely upon Gregory's review, and have to turn to the publications he deals with—and some others, with which he does not deal. We will see immediately that they deserve a clear and patient analysis.

The Italian polemic starts, as a matter of fact, not with Stefano degli Angeli's *Considerationi*, but with some remarks of John Alphons Borelli, which he inserted in his treatise *De vi percussionis*, published in the same year 1667 as the work of the Paduan professor, yet, obviously, some months earlier than this latter. It is, by the way, rather strange that Gregory, who must have known the book of Borelli,[143] does not mention at all these rather pertinent and penetrating remarks—all the more strange since, as we shall see, Stefano de Angeli discusses them at length. It is possible, of course, that Gregory did not want to mention Borelli because of their rivalry in the elaboration of the theory of percussion.[144] Yet this be as it may. In any case, Borelli's analysis of the trajectory of the falling body, though it is, of course, erroneous, is the best ever made before Hooke and Newton. He is the only one who succeeds in disentangling the purely mathematical point of view from the physical. He is, too, the only man before Hooke, who is not dominated and befogged by the traditional conception according to which, whether the Earth moves or stands still, a heavy body has, in any case, to move to the center of the Earth on a perfectly straight line though, of course, he still believes that it will ultimately arrive there. Thus, on the pages 108–109 of the *De Vi Percussionis*, we read the following passage, where, taking into account the theories of Galileo, Fermat, and Mersenne—though not naming them—Borelli writes:[145]

---

[140] To analyze Gregory's theory of the force of percussion, or to evaluate its interest for the history of dynamics and its struggles for a correct measurement of the *vis percussionis*, lie beyond the scope of the present paper. The meaning of this theory, as an argument against Riccioli, consists in the suggestion that, since the trajectory of the falling body, *in hypothesi terrae motae*, would be a curved line (a circle), it would arrive at the surface of the earth at different angles to this surface, and that, as the angles increase more and more, the power of the percussion will increase too.

This reasoning is so faulty that one is astonished to find it adopted by Gregory, or, better to say, that one would be astonished if one had not already seen rather numerous examples of the difficulty of understanding the fundamental concepts of the new dynamics and of the, even greater, difficulty of their application to the physical world. On James Gregory, cf. Herbert Western Turnbull, *James Gregory*, London, G. Bell, 1939.

[141] Needless to say, Descartes never made such a stupid experiment. The story is told by Stefano degli Angeli in his *Seconde Considerationi*, 100.

[142] The paper of James Gregory was read before the Royal Society on June 15, 1668.

[143] James Gregory studied in Padua from 1664 to 1668, and it is in Padua that he published the first (in 1667) and the second editions (in 1668) of his *Vera circuli et parabolae quadratura*.

[144] Whereas according to Gregory the force of percussion is proportional to the sine of the angle of incidence, it is, according to Borelli, proportional to the sine of the complement of this angle. Cf. *supra*, p. 357 and J. A. Borelli *De Vi Percussionis*, prop. XXXX, p. 83, Bononiae, 1667.

[145] As the passage is quite particularly important, I shall quote it in the original: J. A. Borellus, *De Vi Percussionis*, 107–108: "Occasio postulat, ut aliquid innuamus de motu mixto ex transversali circulari aequabili, et ex perpendiculari descensivo uniformiter accelerato versus centrum ejusdem circuli, qui motus neque per circuli peripheriam fit, neque per parabolam, neque per helicam peculiarum ejus naturae, quam aliqui recentiores putarunt. Sit circulus AB cujus centrum C, moveatur vero corpus A impetu transversali circulari per AB, sed aequabili, descendat simul versus centrum C motu uniformiter accelerato, efficiet quidem transitum curvum CGH, quem circu-

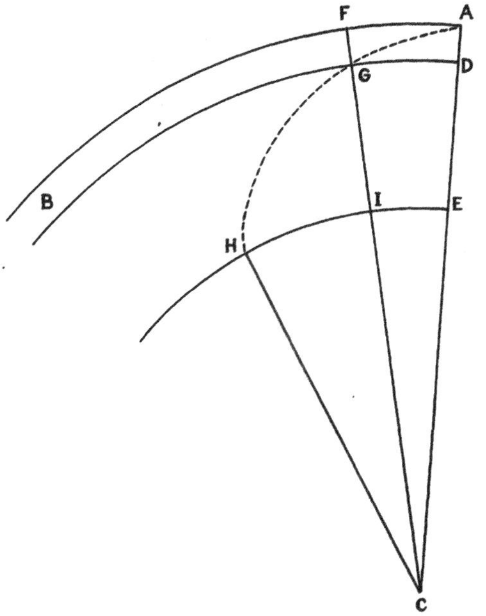

FIG. 16.

It is necessary now to say something about the motion compounded by a uniform circular and a perpendicular descending [one] uniformly accelerated toward the center of the circle; this motion proceeds neither on the periphery of a circle, nor on the parabola, nor on a certain peculiar spiral, as some of the more recent [writers] have believed.

Let there be a circle AB [fig. 16], the center of which be C, and let the body A move on AB with a circular

larem non esse praecise facile demonstrari potest, quia descensus haberent, quam habent sinus versi semissum arcuum excursorum, AH et AG et ideo haberent proportionem minorem quam duplicatam temporum, quod est falsum; similiter non esse parabolam perspicuum est; remanet postrema sententia existimantium esse helicam non difformem ab Archimedea, nisi tantummodo in motu accelerato versus centrum, putant igitur temporibus aequalibus circa centrum C pertransiri angulos aequales ACG, et GCH, quibus aequalibus temporibus percurrunt spatia AD et DH quae sunt in ratione unius ad 3. Sed praedicti auctores non animadverterunt se in hypothesi assumpta non persistere, supponunt // enim eodem impetu transversali mobile A moveri et siquidem grave A perpetuo permanerit in peripheria AB profecto temporibus aequalibus percurreret spatia aequalia subtendentia angulos aequales ad centrum, at quia mobile perducitur ad circumferentias circulorum continenter decrescentium fit ut spatia illa inter se aequalia quae ab impetu perseverante ejusdem roboris percurruntur subtendunt successive angulos majores ad centrum; quare si primo tempore mobile excurrit spatium DG secundo tempore ei aequali percurret spatium IH aequali ipsi DG et quia hujusmodi spatia aequalia mensurantur non in eodem, sed in divertis circulis inaequalibus fit ut angulus ACG minor sit angulo GCH, et proinde anguli praedicti successive crescunt prout distantia a centro G diminuuntur non tamen eadem proportione ut facile ostendi posset, unde constat curvam lineam AGH non esse regularem."

transverse and uniform *impetus*, but at the same time descend toward the center C with a uniformly accelerated motion; it will certainly describe a curved path AGH, which can easily be demonstrated not to be exactly circular; for the descents will be as the versines of the halves of the arcs AH and AG taken separately and therefore they will be in smaller proportion than that of the square of the times, which is false; likewise, it is clear that [this motion] will not describe a parabola; [there] remains [to be considered] the last opinion, of those who believe it to describe a spiral not differing from that of Archimedes, but in respect to the accelerated motion towards the center; they believe therefore that in equal times [this motion] will describe around the center C equal angles ACG and GCH, and that in these equal times [the thus moving body] will traverse the spaces AD and DE which are [to each other] in the ratio of one to three.

But the above-mentioned authors do not recognize that they do not persevere [in holding] the assumed hypothesis; for they have admitted that the moving body A moved with the same transverse *impetus*; thus if the heavy body A remained perpetually on the periphery AB it is clear that in equal times it would traverse equal spaces subtending equal angles to the center; but as the body is moved on circumferences of ever diminishing circles, it results that these equal spaces that are traversed by the *impetus*, which perseveres in the same force, will subtend successively greater angles to the center; therefore, if in the first [unit of] time the body traversed the space DG, in the second [unit of] time, equal to it, it will traverse [the space] HI equal to DG, and as these equal spaces are measured not by the same, but by different and unequal circles, it follows that the angle ACG will be smaller than the angle GCH, and therefore these [subtended] angles will increase as the distance from the center diminishes; not in the same proportion, however, as can be easily shown; from which it is clear that the curved line AGH is not regular.

Borelli is perfectly right in stating that the "above-mentioned authors," that is Fermat and Mersenne, do not persevere in holding their initial hypothesis; in other words that they forget they are dealing with a physical body endowed with a certain, determinate *impetus* or degree of velocity, which degree of velocity does not change with the said body's removal or approach to the center it circumgyrates. They do not see that this body cannot be treated in the same way as can, and must, be treated a geometrical point running up or down on a uniformly rotating radius. In the former case the body maintains its *linear* velocity and therefore traverses on all the circles on which it moves (or, more exactly, would traverse on all these circles, if it stayed on them) the same linear distance: it moves accordingly with an "unequal" angular speed. In the latter case the point circumgyrates with a constant angular velocity, which means a constantly changing *linear* one, or, to use Borelli's terminology, a constantly changing *impetus*.

Borelli continues (pp. 110 sq.) to explain that the force of an "oblique" percussion is to be measured not by the *impetus* along the "oblique" path, but only by that on the perpendicular, which, he adds, enables us to dispose of the argument proposed some time ago by a celebrated author (Borelli does not name Riccioli) according to whom, if the Earth moved, the motion of a body falling from the top of a tower would be uniform, and, therefore, the force of percussion would not increase, at least not perceptibly, with the increase of the altitude of the fall. The celebrated author forgets, Borelli explains, that the point or plane of impact does not stand still in world-space, but is transported together with the tower and the body falling from its top.

The contradiction implied in the "spiral" conception of the fall seems obvious enough. Yet not only was it never noticed before Borelli, but even after it was revealed by him, it remained, as we shall presently see, beyond the grasp of most of his contemporaries. Strangely enough, the R. P. Giambattista Riccioli, though he was never able to understand the Galilean conception of motion, grasped the difference, whereas Stefano degli Angeli, who, at first glance seems to have understood it quite well, did not.

But, before proceeding to the study of Angeli's criticism of Borelli, I would like to point out the curious inability of Borelli himself to apply correctly the principles of the new mechanics to the analysis of the fall. This inability is all the more curious because he had applied these principles himself, with a much better result, to the analysis of the celestial movements in his *Theorica Medicearum Planetarum,*[146] published, or in any case written, a couple of years before the *De Vi Percussionis*.

Indeed, in this latter work, Borelli, as we have seen, speaks about a uniform circular *impetus*, forgetting, as it were, that in the new mechanics, a uniform *impetus* is always a rectilinear one,[147] whereas in the former work he had freed himself from the obsession of the circularity to the point of introducing—for the first time in the history of astronomy—centrifugal forces in the mechanics of the heavens.[148] Moreover, studying the case—perfectly analogous to the one he is dealing with in the *De Vi Percussionis*—of the motion of the celestial bodies tending toward, and gyrating around, the Sun, he deduces from the "compounding" of these two motions the elliptic trajectory of the planets.

Why did he not transpose his reasoning to the case of the body falling from the tower? It is quite improbable that, in the meantime, he had recognized the erroneousness of his deduction of the elliptical path of the planets; if so he would have stated it. It is obvious that, even for him, the analogy—so evident for us—did not exist, and that the case of a heavy ball falling vertically (at least *ad sensum*) from the Torre degli Asinelli had nothing to do with that of the planets revolving around the sun: the planets had not to

arrive at the center of the Universe, but the heavy body had to reach the center of the Earth.

Thus he did not perceive the analogy between the celestial and the terrestrial motions, and his analysis of the descent of the bodies *in hypothesi terrae motae* became rather faulty. He did recognize, nevertheless —and this is not a small merit—the inevitable consequence of the new conception of motion (which, as we have seen, Galileo failed to grasp), namely, that if the Earth moved, the falling ball would not move parallel to the tower, nor would it fall at its foot. Neither would it, as supposed by the traditional theory, still held by men like Riccioli, lag behind and fall to the west. It would, quite on the contrary, outrun it and *fall to the east* of this tower [149]—a conclusion so glaringly opposed to common sense and common experience that Borelli could not accept it himself and that it became the basis of Stefano degli Angeli's counter attack on Borelli.

VIII. S. DEGLI ANGELI CONTRA BORELLI AND RICCIOLI

Stefano degli Angeli's *Considerationi,* written as a frank imitation of the famous *Dialogue* of Galileo, in the form of two dialogues between Count Leszczinsky,[149a] an otherwise unknown Venetian called Ofreddi, and Stefano degli Angeli himself, introduced under the transparent designation of the "Mathematician," [150] are chiefly devoted to the criticism of the arguments, old and new, presented by the R. P. Giambattista Riccioli against the diurnal motion of the Earth. Stefano degli Angeli— obviously a Copernican—takes in his work a safe position: he firmly believes in the immobility of the Earth and asserts it as a divinely revealed truth—which, therefore, does not need any rational or experiential demonstration—and he criticizes not the thesis, but the reasonings and proofs of Riccioli. Like Galileo in his *Dialogue on the two greatest world systems,* he endeavors, by multiplying carefully analyzed examples, to familiarize the reader with the fundamental concept

---

[146] G. A. Borellus, *Theorica Medicearum Planetarum ex causis physicis deducta,* Bononiae, 1666. On Borelli's celestial mechanics *cf.* my paper, La mécanique céleste de Borelli, *Revue d'Histoire des Sciences* 6: 100–138, 1952.

[147] In the modern—Galilean and Newtonian—dynamics, the circular motion is not a uniform, but an accelerated one.

[148] *Cf.* my paper La gravitation universelle, de Kepler à Newton, *Archives Internationales d'Histoire des Sciences;* 1951.

[149] It is worth noticing, that though he recognized this consequence of his theory, he could not believe it and therefore did not announce it to the world; at least not in print. The *De Vi Percussionis* is mute about it. *Cf. infra,* p. 374.

[149a] Probably Bohuslav Leszczinsky (1633/4-1691) who, with his younger brother, Jan, was in Padua in the sixteen sixties. Very little is known about Jan who died in 1668. As for Bohuslav—*Reverendissimus D.D. Bohuslaus Comes de Lesno Leszczynski Praepositus Procensis, Canonicus Cracoviensis,* he was elected Counsellor of the "Polish Nation" of the University of Padua; in 1679 Queen Eleonora made him her chancellor. His nephew, Stanislaus, became King of Poland in 1704.

[150] Stefano degli Angeli (1623-1679?), Jesuate—and not Jesuit, as in Poggendorff's *Biographisch-Literarisches Handwörterbuch zur Geschichte der exacten Wissenschaften* 1: 46, Leipzig, 1863—was professor of mathematics in Padua and wrote a number of works on the quadrature of geometrical curves, and also two books on spirals.

of the new dynamics, the Galilean relativity of motion.[151] But he does not restrict himself to this purely negative task. He presents us with a positive solution to the—hypothetical—case of a heavy body descending *ex sublime* on a moving Earth.

Angeli's criticism of Riccioli has been sufficiently dealt with by James Gregory and I need not dwell upon it any longer. It is to the positive solution (passed over by Gregory) that we now will turn our attention.

This solution, as a matter of fact, is exactly that of Fermat. Stefano degli Angeli does not quote the great French mathematician,[152] but presents it as his own solution of the question, raised by Count Leszczinsky, about the true nature of the curve described by a falling body in its motion to the center of the Earth:[153]

MATT. Let us suppose that the Earth moves either only with the diurnal motion, or with the annual too. Either on the Equator, or on another parallel. I have only thought out what will happen on the Equator assuming the diurnal motion alone; because this motion, continued up to the center, would be a spiral line, which I would be able to describe by points in the same way in which geometers describe the conical sections, or other curved lines. Moreover, I know exactly the proportion of the spiral space to the sector of the circle which includes it.

The Count being naturally eager to learn the demonstration, Stefano degli Angeli (the mathematician) pursues:[154]

Well, I shall oblige you quickly.
We assume, *first*, to be established that the heavy body descends in such a way that the spaces traversed are as the squares of the times. *Second*, [that the straight line] AC made up by the semidiameter of the Earth and the Tower of the Asinelli has 25,870,240 roman palms [as it is admitted by Riccioli]. Third, [that] in one second of the hour the heavy body descends 15 feet. From these suppositions we shall investigate in the following way the amount of time that the body will consume in reaching the center. The proportion which 15 feet, that is the distance

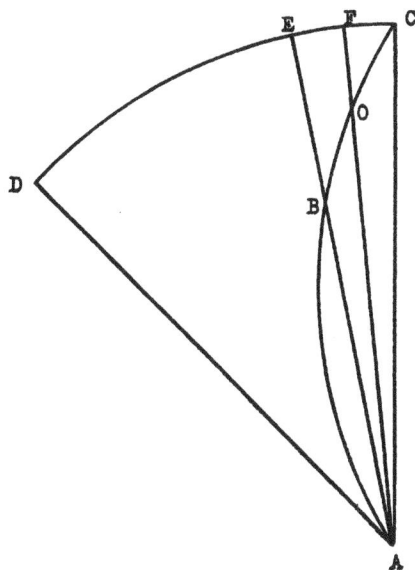

FIG. 17.

traversed in one second, have to 25,870,240, is the same as that which the square of one second has to 1,724,682⅗ [seconds], that is to the square of the time of the whole descent to the center. Of which the nearest square root, 1303, will be the number of the seconds of the hour, which it will consume in arriving there. This will be 21 minutes and 43 seconds *ad proximum*. Which, if one converts them into degrees of the Equator, will be, once more *ad proximum*, 5 degrees, 25 minutes, 45 seconds. If then we consider the sector of the Equator DAC of which the arc DC be 6° 25′ 45″,[155] the spiral line CBA will be included in it. Therefore, if we wanted to describe this line by points, taking for instance an arc of the Equator of 4″, CF and drawing [the semidiameter] AF; and if on this we took FO equal to 15 feet, the spiral would pass through the point O. In the same way, [if we] take [the arc] CE of 8″ and determine on the [semidiameter] AE [the segment] EB, equal to 60 feet, the spiral would pass through B. And in this way [we can determine] the others. Wherefrom it can be argued [that] the spaces of the spiral traversed in equal times will be all the greater as the moving body approaches nearer the center. And, in consequence, as it moves more rapidly on the semidiameter.

The Count, once more, asks the pertinent question: "But how do you know the geometrical proportion of the space CBA to the including sector DAC?" This enables Stefano degli Angeli to explain:[156]

In a work which I hold ready for the press, I investigate two kinds of infinite spirals, taking into consideration the circle, or whatever sector CDA you want, and supposing [that] two motions [are] performed in the same time, the one, [that] of the semidiameter CA, made around the center A on the arc CED, the other, [that] of the point

---

[151] The example of a moving ship, of a ball moving down inside a vertical hollow rod at the same time being horizontally transported by a runner, of a cannon, moving while its ball is shot, etc.

[152] Though it may seem rather improbable that Stefano degli Angeli should ignore Mersenne's *Cogitata Physico-Mathematica* of 1644, all the more as his analysis of the example of the moving cannon conforms strictly to that which Mersenne gives in his *Phenomena ballistica*, yet we must not forget that Angeli has not only been a keen student of the geometry of spirals—he published in 1660 *De infinitorum spiralium spatiorum mensura opusculum geometricum*—but also recognized the particular character of Fermat's spiral and made a study of the inverted spirals in his *De infinitis spiralibus inversis*, published in Padua in 1667. Moreover, when accused by Borelli (*cf. infra* p. 371) of having borrowed his solution from Fermat, S. degli Angeli protests vehemently and asserts that he did not know it as he had not studied, but only cursorily looked through Mersenne's work (*cf. infra* p. 377).

[153] *Considerationi sopra la forza di alcune ragioni fisico mattematiche, addote dal M.R.P. Gio. Battista Riccioli, espresse in dve Dialoghi* da F. Stefano degli Angeli, 26, Venetia. 1667.

[154] *Ibid.*, 28.

[155] Obviously a misprint: it should be 5°25′45″.

[156] *Ibid.*, 29.

C [moving] from [the point] *A* on the [semidiameter] *CA*, and proportioned in such a way that the motion of the semidiameter on the arc be uniform, and that of the point *C*, on the [semidiameter] *CA* be uniform, or accelerated, according to whatever power of the times; or *vice versa*, this one uniform, and that one either uniform or accelerated. And I have [determined] geometrically the proportion of all these spaces to the [corresponding] sectors. The demonstrations depend on many things; thus you will be able to see them only when I publish them.[157] Meanwhile, it will be sufficient to know, that in our case, in which the described spiral will be the second of the first order, the sector [*DAC*] to the space [*CBA*] will have the proportion of fifteen to eight.

So far, so good. But the Count has recently seen a book of Giovanni Alfonso Borelli in which this latter denies that the curve in question will be a spiral. Stefano degli Angeli has seen it too, having received it as a kind gift from the great mathematician. He has had no time yet to study it, but the book is at hand. What does it say?[158] "Well," answers the Count, who meanwhile has taken the book and opened it on the relevant page, "it says just that, viz. that this line will be neither a circle, nor a parabola, nor any kind of spiral." Stefano degli Angeli agrees, of course, that it will not be a circle or a parabola, but why not a spiral? "Because, quotes the Count:[159] 'Let there be a circle *AB*, the center of which be *C*, and let the body *A* move on *AB* with a uniform circular transverse *impetus*, but at the same time descend toward the center *C* with a uniformly accelerated motion; it will certainly describe a curved path *AGH*.' He says that it will be neither a circumference, nor a parabola, nor a spiral." The discussion then starts:

MATT. Till now, as for the substance of this his explanation, I do not see any difference from my own. And he says that it will not be a spiral?

CONT. Indeed, he says so. Thus towards the end he concludes: "from which it is clear that the curved line *AGH* is not regular," but he does not give any reason.

MATT. Read it please [more] distinctly.

CONT. I shall begin higher: "There remains the last opinion of those who believe it to describe a spiral not different from that of Archimedes, but in respect of the accelerated motion toward the center; they believe there-fore that in equal times [this motion] will describe around the center *C* equal angles *ACG* and *GCH* and that in these equal times [the thus moving body] will traverse the spaces *AD* and *DE* which are [to each other] in the ratio of 1 to 3."

MATT. I hold it to be certain.

CONT. If it is so, then you will be of the number of those of whom he says: "But the above mentioned authors do not recognize that they do not persevere [in holding] the assumed hypothesis."

MATT. But why?

CONT. For they have admitted that the moving body *A* moved with the same transverse *impetus*; thus if the heavy body *A* remained perpetually on the periphery *AB* it is clear that in equal times it would traverse equal spaces subtending equal angles to the center.

MATT. If it is so, then the Author does not seem to persist in his supposition either. Because indeed he says *supra*: "Let the body *A* move on *AB* with a uniform transverse circular *impetus*." Now the moving body is not on the arc *AB*; except in one point at the beginning of the motion, in the progress of which it is always elsewhere. And thus it does not move on this [arc], as it is assumed. But please, let me see what he says. I believe that I have understood. I think that we disagree on the principles, because I believe that those who have said that this line is a spiral wanted to say (or at least this is the way in which, according to what I have declared *supra*, I myself want to be understood) that the motion of the moving body is compounded by two motions, that is, by the uniform [motion] of the semidiameter *CA*, fixed at the center *C*, along the arc *AB*, which motion is in the moving body by participation, as in a body which is always on the semidiameter which moves with this motion; and of that of the heavy body *A* [itself] moving downward along the semidiameter, [and] accelerated in such a way that the spaces traversed are as the squares of the times; which motion is proper to the moving body.[160] And this first motion would be found in Nature provided the Earth moved with the diurnal motion only; because it would be described by the heavy body falling naturally on the plane of the Equator. The said motion would be uniform in respect to the arc *AFB* because in equal times it will traverse equal arcs and by these [arcs] would be subtended equal angles to the center: *vice versa*, equal arcs would be passed in equal times. And though it be true, that in every point of the curve *AGH* the heavy body would have a different circular *impetus*, because all the points of the semidiameter move with motions of different speed, that is, more slowly as they approach the center. Nevertheless, this is true for the heavy body wherever it be, and is not referred to the arc *AFB*. As I have said, I do not believe that those who have said that this motion will be along a spiral have ever understood this motion to be compounded in any other way. And I infer from what the Author says *supra*: "there remains the last opinion of those who believe it to describe a spiral not differing from that of Archimedes but with respect to the accelerated motion toward the center." Who, indeed, does not know that, of the two motions concurring in the description of this [spiral], one is the proper [motion] of the point along the semidiameter toward the circumference, the other that of the semidiameter fixed at the center of the circumference and carrying with it this very point? Now I understand that the line which we call spiral is described in this way,

[157] As I have just said, Stefano degli Angeli published his book in the same year, 1667, as the *Considerationi*, to which he refers in the preface; cf. Stephano de Angelis, *De infinitis spiralibus inversis*, Padova, 1667, preface: "Jam anni quadrans evolavit, ex quo discursus quidam nostri italice conscripti prodiere; in quibus rationes aliquot physico-mathematicas contra Coperniceum systema a viro eximio, Ricciolio excogitatas, ad trutinam revocavimus. Affirmabamus tunc bina genera infinitarum spiralium a nobis fuisse contemplata; quarum una illa foret, quae a gravi naturaliter cadente in plano aequatoris describeretur si Tellus, ex falsa hypothesi, motu diurno duntaxat moveretur, et grave deorsum latum taliter suum concitaret motum, ut spacia ab ipso peracta, forent ad invicem ut ipsa quadrata temporum."

[158] *Ibid.*, 29; as a matter of fact, as Angeli reports in his *Terze Considerationi* (cf. *infra*, p. 371) Borelli had criticized the "spiral" solution even before the publication of his *De Vi Percussionis*.

[159] *Ibid.*, 30; cf. *supra*, p. 359 and fig. 16.

[160] The distinction, made by Angeli, between the "proper" motion of the heavy body toward the center of the Earth, and of its motion along the arc *AB*, which it performs only in virtue of its participation to the motion of the terrestrial globe is the key to all the geodynamical conceptions—and errors—of Angeli.

and I believe that the others have understood it the same way. And if the Author understands it differently, as indeed he seems to me to do, I do not think that he contradicts the others who have made suppositions different from his [own]. The Author then should be understood in this way: "Let the body A move on AB with a circular transversal but uniform *impetus.*" Not meaning that "the moving body would move uniformly on AB but that it should move with the *impetus* which it would have if it would move on AB." [161]

And I believe that he understands it thus, because, after having said that the moving body in equal times would describe equal arcs if it remained on the arc AB, he adds: "but as the body is moved on circumferences of ever diminishing circles, it results that these equal spaces, that are traversed by the *impetus* which perseveres in the same force, will subtend successively greater angles to the center; therefore, if in the first [unit of] time the body traversed the space DG, in the second [unit of] time, equal to it, it will traverse the space IH equal to DG, and as these equal spaces are measured not by the same, but by different angles, it follows that the angle ACG will be smaller than the angle GCH, etc." In this case the traversed space would not be the spiral line described by us, but another [one]. Yet it would be nevertheless true, that not this one, but our [spiral] would be described by the heavy body if the Earth moved with the diurnal motion only in the plane of the Equator; because the heavy body, in whatever place it be, would have that circular speed, which is required by its distance from the center. As for his [Borelli's] composition of motions, I do not believe there could be any such in Nature.

Stefano degli Angeli is a very good example of something that we encountered all along our investigations, namely of the tremendous power of mental, or intellectual, inertia and of the very slow and gradual way in which even superior minds succeed in liberating themselves from the traditional and habitual *idola tribus.* He is, as we have seen, much farther on this way than Riccioli. He has grasped the meaning of the Galilean relativity of motion: he does not believe any more—as Riccioli still does—in the *impetus* as a cause or force impressed on, or inherent in, the body, force, or cause which produces motion as its effect and which, therefore, is not indifferent to the presence or absence, in this very body, of another *impetus* of the same, or of a different kind. For him, as for Borelli, *impetus* is the same thing as motion, and each *impetus,* therefore, is independent of and compatible with each other. He understands perfectly well the exact meaning of Borelli's theory—so well that he is able to draw therefrom its hidden implication. Yet he rejects it for the rather curious reason that Borelli does not take into account the difference between the "proper *impetus*"—or motion—of the body toward the center of the Earth, and its motion "by participation" only, which it has from

the moving Earth and in which it does not move *itself* but is merely carried around by the Earth.

In other words, for him the natural motion of the heavy body downward, toward the center of the Earth, still possesses a unique and privileged character and is qualitatively distinct from any other kind of non-natural motion. [162] This seems to be quite particularly true with respect to the transverse, circular motion that the body accomplishes only because of the motion of the Earth. It is not its *own* motion; it receives it only *from outside.* The principles of the new dynamics are obviously too loosely fixed in Stefano degli Angeli's mind to enable him to think accordingly. Thus, the body which participates in the rotation of the Earth is not conceived by him as sharing this movement, but only as submitting to it. The body does not move *with* the Earth; it is moved *by* the Earth, and this because it is, physically, attached to it. [163]

Thus the physical and the geometrical intuitions concur in singling out the spiral theory as the only one which is not only geometrically possible but also physically true—in contradistinction to that of Borelli which, curiously enough, appears to Angeli as a purely mathematical construction without any basis in reality.

Angeli's conceptions are expressed very clearly in another passage of the *Considerationi,* directed against both Riccioli and Borelli. Starting with the discussion of the famous Tychonian argument (revived by Riccioli), according to which the very rapid diurnal motion of the Earth would impede, if not completely hinder, the so much slower downward movement of the heavy body, because of the incompatibility of the two *impetus* in the same mobile—an incompatibility which Angeli denies, asserting on the contrary their perfect interindependence [164]—he continues, [165]

The heavy body that would fall along the perpendicular BA [fig. 18], would, with this same perpendicular, be carried in a circle by the diurnal revolution; that is, it would not be carried to the other perpendiculars HQ, IF, etc., but it is the same perpendicular BA, which we have supposed to be the Torre degli Asinelli and which at first occupied the site BA, that would then be in the sites HQ, IF, etc., in succession. Wherefrom [it follows that] the falling body would never be separated from the Tower BA. Though this [Tower] would be, in turn, carried in circle by the diurnal revolution, the eye placed on the Earth and carried by the same movement, would not see anything but the descent along the perpendicular. But the eye placed

---

[161] *Ibid.,* 32. Stefano degli Angeli wants to stress that, the transverse *impetus* being in the body only by participation, it cannot remain the same when the said body moves on the radius of the rotating circle toward its center; it would do it if the transverse *impetus* were proper to the body, which it is not. See fig. 16.

[162] We must not forget, however, that in a time in which the difference between *mass* and *weight* was not yet recognized, this latter appeared as something not only essential to, but constitutive of, a physical body. But weight implies, or even means, motion downward.

[163] Even if not mechanically connected with the Earth, the heavy body is, nevertheless, seen by Angeli as bound to it; it is never free.

[164] *Ibid.,* 99: "Questo non è impossibile, anzi necessario. Ogn'uno questi principij independemente dall'altro nel medemo tempo cacia il mobile per diverse strade; la revoluzion diurna la porta in giro; la gravità la porta allo ingiù."

[165] *Ibid.,* 100–109.

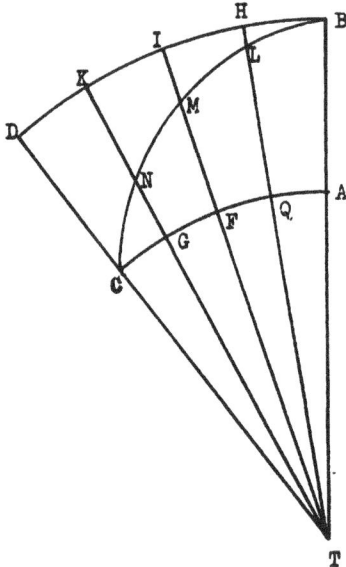

FIG. 18

outside of the Earth would see all the [motions] which are, physically and really, performed. Therefore it would see the heavy body B moving toward A on BA. It would [also] see the BA not remaining in the same site, but being transported successively in the sites HQ, IF, etc. And it would see the heavy body B, in these different sites, having different positions with respect to the Tower BA, that is, being in B, L, M, etc. All these points will represent a spiral. From there [166] all these things would be seen by the eye. They would happen in this way and not in another; nor is the accomplishment of these two different motions, that is of the descending and of the circular ones in the same time, impossible, but necessary, physical, and real. It results therefrom that, the process occurring in this way, all the others which the Author [Riccioli] endeavors to condemn are impossible; therefore I subscribe to what he says.

CONT. Please be not so precipitous in conceding anything; indeed, in the same §, somewhat further, he says that it is impossible for the same moving body, if it were not redoubled by God's power, to perform at the same time both these motions, that is the [motion] downward and the circular [one]; just as it is impossible for the same moving body to move at the same time in opposite directions of the world, that is, for instance, along the Equator to the Orient and on the Meridian toward the Pole; but if it is carried along a median way, it has to be conducted to the goal by an oblique path, so as to move neither simply toward the Orient nor simply toward the Pole, but to participate in the one and in the other [motions], as for instance [when traveling] from Greece to the Levant etc.

MATT. When things, that are taken to present a similitude, harmonize in all the necessary conditions, they represent *quasi* the same thing. Thus these two motions performed on the Meridian toward the Pole, and [on the

[166] *Onde* = from the point outside of the Earth.

Equator] toward the Orient, duly adjusted, can make us understand *suo modo*, the motion of the heavy body in the Copernican hypothesis. Let us imagine, then, for greater facility, on the right sphere,[167] an ant [placed] on the intersection of the Meridian and the Equator; let this [ant] be carried by its own appetite along the Meridian toward the Arctic Pole. Let at the same time the Meridian of the intersection, with its horizon, be carried toward the Orient. In this case the ant will be moved by two real motions accomplished at the same time, that is by its own [motion] always on the same Meridian toward the Pole, to which, at the end of the motion, it would have approached in proportion to the speed; and [by the motion] it performed toward the Orient [having been] carried by the Meridian. And from these two physical and real motions will be born the resulting transverse, for instance, from Greece to the Levant.

Perfectly true—an ant running on the meridian of a rotating sphere will certainly be seen by an observer placed outside of it as performing at the same time two different motions, that is (a) as approaching the pole of the sphere along the meridian, and (b) as transported by this sphere transversely with an ever diminishing speed, a speed equal to the linear velocity of the point occupied by it on the meridian. It will, therefore, describe in the absolute space a rather complicated curve, the plane projection of which would correspond to the spiral it would describe if it were running not on the meridian but on the radius of the sphere. The analogy thus seems to be just of the kind requested and to justify an appeal to the authority of Riccioli himself:[168]

OFREDDI. I remember a very subtle reasoning, which the same Author, with great perspicacity, has proposed to the Copernicans (in the book IX of the *Almagest* at the end of the chap. 9 § 5) and which seems to me [to] show very clearly [that] both these motions, that is, [the motion] on the straight line and [on] the circular one, made at the same time by the same heavy body, are possible also according to this very Author, who says these words: "We have ourselves suggested to the Copernicans an easier way to experiment the effect of the motion of the same body, produced by two principles, of which the one moves on the straight line and not uniformly, whereas the other on the circle, and uniformly."[169] The way is the following one: AC [fig. 19] is a beam perforated along the length AC by a small channel [and] placed on stagnant water;[170] of this beam the end A is held firmly as a center, and the other end C is conducted in a circle, with a uniform motion, through G to D; let there be in C a small globe, to which

[167] *Sfera retta* = the sphere rotating from West to East.

[168] *Ibid.*, 102–109.

[169] The experiment imagined by Riccioli endeavors to show that the circular motion of the body hinders its motion toward the center. Riccioli is perfectly right except for one thing: he forgets that this hindrance is occasioned by the centrifugal force of the rotation.

[170] This "experiment" of Riccioli could very well be the source of inspiration of the famous "experiments" of Borelli, described by the latter in the *Theorica Medicearum Planetarum*. The only—but very important—difference is that in Borelli's experiment the beam is inclined, and that it is just the action of the centrifugal force—and not of the circular motion *as such*, that he wants to demonstrate. *Cf.* my paper quoted *supra*, n. 146.

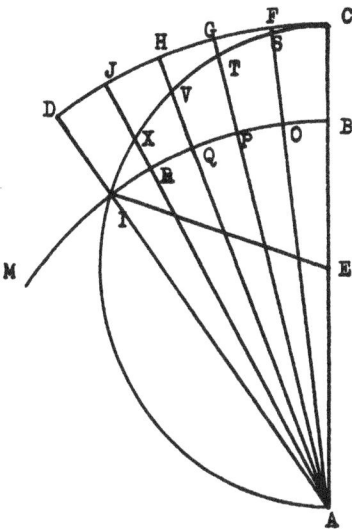

FIG. 19

there will be attached a string, which, on the other extremity, would be joined to a great weight, able to pull it
through the small channel.

MATT. No more, no more Sign. Ofreddi: I understand
it all. A certain and a very beautiful way, and [which]
explains exactly the description of our spiral. The weight
would pull the small globe always through the same small
channel physically and really. The hand physically and
really would conduct the channel and the small globe in a
circle. And because of these two physical and real motions,
a spiral would be described in the water.

Once more, the classical examples of the interindependence of motions are marshalled against Riccioli.
The spiral theory of fall emerges victorious over all
objections, and, to close the discussion, Stefano degli
Angeli shoots, not at Riccioli, but at Borelli, a quite
particularly damaging Parthian arrow.[171]

OFREDDI. I don't know whether Sign. Borelli has
noticed that, if his theory were true, and the heavy body A
[fig. 16] moved always with the same speed which it had
in A, it would have as consequence something to pull P.
Riccioli seems to me to have attempted in vain; that is,
he would have demonstrated by a most certain and more
than physico-mathematical reason the immobility of the
Earth.

MATT. If it is so, it is necessary to get busy to prove
this his supposition, because it will acquire a great merit.
But is it true that this principle being established, the conclusion [of the immobility of the Earth] would infallibly
follow therefrom?

OFR. I believe it to be most evident. But tell me please:
if I should demonstrate that, this principle and the diurnal
motion of the Earth being supposed, the heavy body falling
from the Torre degli Asinelli will necessarily have to
deviate from the perpendicular to the horizon, and to

[171] Ibid., 116 sq.

precede it to the Orient, will I not have accomplished my
proposal?

MATT. Most certainly; because as it is most true that
it [the body] does not deviate from the Tower and that
our eyes see it always near it, it would be necessary to say
that the Earth does not move. But, in grace, do demonstrate this your conclusion.

OFR. Let us repeat his words already examined in the
Dial. I p. 34: [172] "but as the body is moved on circumferences of ever diminishing circles, it results that these equal
spaces that are traversed by the *impetus* which perseveres
in the same force will subtend successively greater angles
to the center; therefore, if in the first [unit of] time the
body traversed the space DG, in the second [unit of] time,
equal to it, it will traverse [the space] IH equal to DG,
and as these equal spaces are measured not by the same,
but by different and unequal circles, it follows that the
angle ACG will be smaller than the angle GCH."

MATT. Do not say anything else, Sign. Ofreddi, as I
believe I understand what you want to say. And it seems
to me [that] you are saying the truth. In the first [unit of]
time the moving body has traversed the space AG, of which
the arc, DG, subtends at the center the angle DCG. In
the second [unit of] time, equal to the first, the moving
body would traverse the space GH, subtending the angle
ICH, greater than GCD; but the perpendicular to the
diurnal revolution in this second [unit of] time, equal to
the first, would subtend at the center an angle equal to
DCG and smaller than GCH. Accordingly it would be between the FC, CH. But the moving body is supposed [to
be] in H. Consequently it would appear outside of the
perpendicular and to the Orient of it.

IX. M. MANFREDI CONTRA
S. DEGLI ANGELI

Quite naturally, Riccioli was roused by the unex-
pected, somewhat disrespectful and sometimes flippant
attack of Stefano degli Angeli (Stefano degli Angeli
says, for instance, that he has not yet perused the
Astronomia Reformata because it is too thick, and much
too expensive) upon his most prized possession, a new
argument for the immobility of the Earth. Using the
pen or the name of Michele Manfredi,[173] Riccioli sent

[172] Cf. *supra*, n. 145, the text of Borelli.
[173] In his preface (*Al Lettore*), Michele Manfredi tells us
that the R. P. J. B. Riccioli had decided not to reply to Angeli,
(a) because he considered it below his dignity, (b) because
his answer, necessarily, would be above the heads of those
people who have been impressed by the considerations of Angeli
and besides, being already seventy years old he could not allow
himself to lose time on such trifles, (c) that there are people
who are such partisans of the Copernican system (though they
hide it) that they are practically impermeable to reason and
go as far as to deny the value of Riccioli's most evident argument.
Yet, being an old and intimate friend and pupil of Riccioli,
and perfectly versed in his works, says Manfredi, he has
persuaded the great astronomer to allow him, Manfredi, to
reply in his place, in order that, on account of the silence of
the truth, the falsity would not be able to claim victory.
Thus far all seems to be straightforward and clear. Yet in
the dedication to Count Francesco Carlo Caprara, Gonfalonier
of the People and of the City of Bologna, Emilio Manolesi,
whom Manfredi entrusted with the dedicating of the book,
writes: "Io non sapeva veramente discernere, s'io doveva restar
obligato alla cortesia del Signor Michele Manfredi per hauer

to the mathematician of Padua what seemed—at least in his opinion—a crushing reply. As a matter of fact, Manfredi's pamphlet does not contain any really new argument; yet it is interesting as it enables us to understand better Riccioli's own position towards the "semi-circular" theory of the fall, and its role in the ensemble of his anti-Copernican arguments; besides, in his replies to Angeli, he scores some really good points.

Stefano degli Angeli had objected to Riccioli's deduction of the semicircular path of the falling body.[174] Riccioli, pointing out that, as a matter of fact, it is not his, but Galileo's deduction, replies,[175]

On the same p. 20 the Dialogist plays the gentleman, by saying: "And let us concede, (in order to be liberal), that his $FS. GT. HV. LX.$ intercepted between the two circumferences,[176] will have the proportion of the squares of the times, yet it cannot be inferred therefrom, that they will be the spaces traversed by the moving body." There is no need to make a show of liberality for something which is owed by justice and necessity. And it is already proved in § 11 that these lines are necessarily in this proportion. But it is not inferred therefrom that they will be the spaces traversed by the moving body in the Copernican system, because in it [this system] the heavy body will not descend along them, but on a curved line described by the terminals $ST\overline{V}X$, which, though in mathematical rigor . . . must not necessarily be circular, still at the beginning of the motion, and in the first four seconds of the hour . . . cannot be greatly different from the circular.[177] Assuming it to be circular, or very nearly circular, as, with Galileo the R. P. Riccioli assumed it in the *Almagestum*, in order to make an argument *ad hominem*,[178] the intervals $FS. GT. HV. LX$ intercepted by the circumferences $CD. BI$ as well as the spaces apparently traversed on the straight line in four seconds of the hour, will correspond [to each other] in the proportion of the squares of the times, which is a

---

rimesso à me l'arbitrio di dedicare à mia voglia quest'Operetta del P. Giambattista Riccioli, Matematico di quella profondita, Astronomo di quella perizia et geometro di quella risoluzione, che per sue immense, e celebratissime fatiche si è fatto hormai noto per tutta l'Europa," etc. As Manolesi says: *quest' Operetta del* and not *questa diffesa*, and as he does not speak about Manfredi as the author of *quest'Operetta*, the implication that this latter has been only a *prête-nom* or a pseudonyme of Riccioli does not seem far fetched. *Cf. supra*, n. 130, 2. In his *Apologia*, indeed, Riccioli says that Manfredi answered in his stead (*cf. infra* p. 393); but that is what he had to say in any case. Michele Manfredi had published, in 1666, *Vindiciae kalendarii gregoriani adversus Franciscum Leveram.* . . . . Bononiae, ex typographia Haeredes V. Benatii, 1666.

[174] *Cf. supra*, Gregory's report, p. 356.

[175] *Argomento fisicomattematico del padre Gio. Battista Riccioli . . . contro il moto diurno della terra Confermato di nuouo con l'occasione della Riposta alle Considerazioni sopra la Forza del detto Argomento etc. Fatte dal M. R. Fr. Stefano degli Angeli. . . .* All'Illustriss, Signore Il Sig. Co., Francesco Carlo Caprara . . . Bologna, 1668, p. 24 § 18.

[176] *Cf.* the drawing on page 338 (fig. 5) or 356 (fig. 14).

[177] It is *ad omnem sensum circularis.*

[178] The structure of the argument is thus the following one: the Copernicans—that is Galileo—assert that the motion of the falling body will be exactly or very nearly circular. The "very nearly" is sufficient for us, because if *they* assume it, *they* have to admit that, on a rotating Earth, the motion of a falling body would be uniform, or nearly so. As it is not, they have to recognize that the Earth does not move.

proof, though not necessary, yet very probable, that in the Copernican system the said moving body would, in the beginning [of its motion], describe an, *ad sensum*, circular line, and consequently would move without a real acceleration, [at least without a] sensible or notable one, compared to that which is required by the increment of the real percussion.

The point raised by Manfredi is really important. Indeed, for his principal argument, the exactly circular character of the trajectory of the falling body is immaterial: it is quite sufficient that this curve be circular only approximately or *ad sensum* (Galileo himself did not assert more than that), and that it be circular at the beginning of the motion, during the first four seconds, that is in the region accessible to our experiments. He is, therefore, perfectly right to insist upon it:[179]

From the end of page 22 and to page 24, the Dialogist makes use of a discourse of P. Riccioli reported in § 14 of chap. 17 of the *Almagestum*, where he proves that with the progress of time, the line [of descent] of the falling [body] described by Galileo, would not be perfectly circular, but would deviate from the circular.[180] But this has nothing to do with the principal argument, in which P. Riccioli restricts himself to the beginning of this motion, comprised in the first four seconds of the hour, whatever the progress of the motion imagined [to continue] till the center of the Earth might be, and he does not need the exactly circular line in order to prove the physical uniformity of the motion in the Copernican system; then, that this line would fall rather inside of the circle described by Galileo, P. Riccioli has demonstrated in the said chap. 17 by the Table of the Sines, based on more certain proportions than are the proportions assumed by the Dialogist in order to prove that it will fall above and outside of the said circle. Nevertheless, because in the beginning of the motion this difference is very small, it does not harm the substantial force of the argument against the Copernican system.

Manfredi is perfectly right, too, in refusing to accept the analogy, admitted or postulated by Angeli, between the motion of an ant on the meridian of a rotating sphere, or of a point on the rotating radius of the Archimedean spiral, and that of the stone falling from the top of the Torre degli Asinelli (though, one must confess, not always for reasons for which *we* would reject this analogy):[181]

There is a great difference between the motion of the said heavy body and that of the point, or of the ant describing the spiral: because the heavy body is not attached, nor is it adherent, to the Tower, and thus it is neither carried in a circle by an extrinsic principle, nor by a motion impressed to it by the Tower, but, according to the Copernicans, it is moved by an intrinsic principle, common to all terrestrial bodies, which, being strong enough to overcome the *conatus* of the gravity inclining [the body] downward, does not allow it to descend really on a per-

---

[179] *Ibid.*, 25, § 19.

[180] Because of the retardation of the motion by the resistance of the air.

[181] *Ibid.*, 56 § 39.

pendicular line; [182] thus, though it finds itself always at the same distance from the perpendicular of the Tower, it does not, therefore, move downward on it, but only finds itself in a certain point of a perpendicular line, corresponding to the perpendicular of the Tower.[183] But the point of which Archimedes speaks, or the ant substituted for it, would not be able to describe a spiral physically and really, if they were not parts included in, or adherent to, the straight line, [and did not] receive the impression of the rotating motion from this [line] as from a moving body really distinct [from them] and [were not] carried, or pulled by it in circle as by an extrinsic principle; in this case, because of the real distinction of the moved body and the moving principle, there is no repugnance between their real motions; just as in virtue of the same distinction it is not repugnant that a moving body [should] move with a contrary motion, nor that [it should] separate itself from the motion of another moving body, which carries it, and that whereas the ship runs the Tramontane,[184] the pilot runs upon it toward the stern to the South.

Let us not, however, mistake the relativity of motion here admitted by Manfredi for the Galilean one that Angeli so persistently, and so unsuccessfully, tried to teach Riccioli. They are strictly incompatible as the next phrases clearly show us :[185]

But it is indeed impossible that the same moving body animated by moving principles, though formally or imaginarily distinct, yet intrinsic [to it, would] by itself move continually on two different lines, even if the motions were not contrary, and that it moved really downward on a perpendicular straight line, and at the same time sideways; but if it has two intrinsic principles and if the one does not entirely impede the other, it is necessary that they combine together into one real motion participating imperfectly in the properties of the one and the other.

Moreover, [continues Manfredi] it could be replied, that the example of the Archimedean spiral is not adequate for the purpose of the Dialogist, because Archimedes assumes the motion of the point on the semidiameter [to be] equal and uniform; and also coordinated with the motion of the semidiameter in such a way, that, when the semidiameter, in its own revolution has described in the plane the whole circle, then the point moved on the semidiameter will pass precisely from one extremity to the other. But for the intention of the Dialogist it would be necessary that the heavy body should move unequally and with a uniform difformity, namely with the proportion of the squares of the times, and thus arrive at the center, if it were not prevented by the impenetrability of the terrestrial globe, long before the completion of the diurnal revolution of the Earth. The mathematician will answer that there are many kinds of spiral lines, and that for his purpose it is sufficient to prove that a moving body can move on a

straight line and at the same time be carried along another and different road. But we shall say that it is not sufficient because to move along one road and, at the same time, to be carried along another is not repugnant in the case of the real distinction between the moved and the moving; but it is indeed repugnant to the same moving body to move in a certain direction and at the same time to move in another; just as it is repugnant that it [should] be carried by the same carrier in one direction and, at the same time, in another.

There can also be the case in which a moving body carrying [another] and really distinct from the carried one, carries it sideways, with such a speed that it slows down the proper speed of the carried; [186] but though this could be controverted, yet there cannot, however, be good reasons to doubt that the velocity of the diurnal motion, intrinsic to the terrestrial bodies, would slow down the speed of the motion downward.

Of the incompatibility, or mutual hindrance, of two motions of the same body, especially when they are produced by two *impetus*, whether natural or not, present *in* this body, Manfredi remains so deeply convinced that, coming back to the famous example of the ball falling from the mast of a—moving or standing still—ship, used by Angeli as an argument against Riccioli, he replies : [187]

Of the example of the ship and of the globe let fall from the top of [its] mast while the ship moves at great speed, alleged by the Copernicans and by the mathematician of Padua, and by Borelli on p. 112,[188] P. Riccioli has not made any use, because it is beside the point; indeed the motion of such a globe downward is occasioned by its intrinsic gravity and [the motion] sideways by the *impetus* impressed on it *ab extrinseco* by the hand attached to, or placed upon, the mast carried by the ship. In this case, the globe, though it would find itself in the points of diverse perpendiculars equidistant to the perpendicular of the mast, would not, therefore, descend on a perpendicular line, but [those] who would be outside the atmosphere would see it descending sideways along a single curved line, resembling the parabolic or the spiral. And if the excess of the transverse *impetus* on the *impetus* of gravity be as great as, in the Copernican system, is the excess of the *impetus* of the diurnal motion on the *impetus* of a ball of chalk of eight ounces, this globe would be seen to descend along an oblique way either uniformly or without a considerable acceleration and the percussion would be a good deal weaker than that which it would make if the ship stood still. But because P. Riccioli has not made any observations of that,[189] and also because of the disparity shown *supra*, he has decided not to use this example subject to a great many diversities and uncertainties.

---

[182] According to Riccioli's interpretation of Copernicus—an interpretation which, by the way, is rather sound—heavy bodies turn around *with* the Earth because they share with it the same intrinsic principle of motion : they are not moved *by* the Earth as a body distinct from them. In order to be able to move them in this way, the said internal principle, or *impetus*, must be strong enough to overcome their natural tendency to go to the center of the Earth.

[183] The motion of the heavy body is perfectly independent of that of the Tower : it does not move along the Tower, and has nothing to do with the perpendicular of the Tower.

[184] Corre à tramontana = runs to the North.

[185] *Ibid.*, 57, § 40.

[186] Thus, even in the case of really distinct principles of motion, two different motions impede each other *as motions*, and the resulting one can share their properties (direction, speed), only in an imperfect manner. In pure mathematics, it is, of course, different. But mathematics is not physics, and does not deal with the real as such.

[187] *Ibid.*, 75 § 49.

[188] Of the *De Vi Percussionis*. The mathematician of Padua is, of course, Stefano degli Angeli.

[189] The only men who really did make this famous experiment were Gassendi, who performed it in 1640 in Marseilles and, perhaps, Thomas Digges, *cf. supra*, p. 349, n. 91.

Riccioli, indeed, did not use the argument, or the example, of the ship; but he did use a number of others, and discussed—or made Manfredi discuss and criticize at length—those used by Stefano degli Angeli. I shall not, however, attempt to study these criticisms here, but will analyze them together with Angeli's counter objections.

X. S. DEGLI ANGELI CONTRA
M. MANFREDI

Stefano degli Angeli could not, of course, remain indifferent to Manfredi's pamphlet. Naturally, he did not like to be treated as an ignorant schoolboy rebelling against the authority of a master who would not even condescend to reply directly, especially since Manfredi, quite openly, accused him of Copernican leanings, if not of being a crypto-Copernican—an accusation that could not easily be disregarded, the more so as it was obviously true.

Stefano degli Angeli does not mention, of course, this last reason for writing his *Seconde Considerationi* [100] He simply tells us that after some hesitation he decided to publish, as a sequel to his first two dialogues, two more dialogues, in order to make himself better understood and to put an end to the useless polemic.

The *Seconde Considerationi* do not add much new material to the debate. Stefano degli Angeli repeats practically the same arguments that he had already presented in his first attack on Riccioli. The chief difference between the two *opuscula* is in the tone and the style. The practically unknown Manfredi [191] could be taken to task in a much rougher way than the celebrated author of the *Almagestum Novum* and the *Astronomia Reformata*,[192] and if, in fact, he was only a mask for Riccioli, so much the worse or so much the better.

As for the objection made by Manfredi against Angeli's own spiral theory of the fall, of a free fall—in contradistinction to that of a fall inside of a hollowed out beam, or to that of the motion of a point along a radius-vector—Angeli seems not to have understood it. He maintains that, in *hypothesi terrae motae*, the falling body will move on a *physical* radius and describe, with the same angular velocity, smaller and smaller circles: [193]

MATT. The diurnal motion carries in a circle the heavy body and the semidiameter; the descending motion carries it toward the center along the same semidiameter which always accompanies it. Because of the motion which carries it downward, the circular one always changes its circumference. From these two [motions] there results in world-space the spiral line participant of the gyrating and of the downward [motion] by which it is compounded. And the percussion is determined by the descending [motion], which being accelerated in proportion to the squares of the times, must produce the diversity of the percussions and of the sounds, as consequence. To say that it does not describe a perfect circle any more than it descends along a perpendicular is nonsense because inasmuch as it descends on the perpendicular physically, it moves always to the right with the same circular motion.[194]

It is clear that Angeli maintains, unchanged, the physico-geometrical conception that we have seen him defend against Borelli in his first *Considerationi*. Thus he concludes: [195]

MATT. In the motion of the heavy body there will be two distinct movables, that is the heavy body and the physical perpendicular. These two will be moved with the same circular motion; and the heavy body will be carried downward by gravity. These principles, the gyrating and the descending, will be really distinct; and for the substance of the motion, it is of small importance whether they are intrinsic or extrinsic, because they are distinct in themselves, as we have said many times.

But what would happen if the new-fangled theories that are becoming increasingly popular and that explain —or explain away—natural gravity by magnetic attraction were true? [196] Ofreddi squarely raises the question: [197]

OFR. But what would happen if the descending principle were not intrinsic to the moved body, and if gravity were nothing else than a magnetic virtue with which the great body of the Earth draws toward itself the other small bodies separated from it, as the larger pieces of magnet draw to themselves the smaller, etc.? One could say, perhaps, that, if somebody could separate a certain part of this our Earth, and carry it very high, it would be outside of the sphere of the activity of the remaining

---

[100] *Seconde considerationi sopra la forza dell'argomento fisicomattematico del M. Rev., P. Gio. Battista Riccioli . . . contro il moto diurno della, terra, Spiegato dal Sig. Michiel Manfredi nelle sue Risposte, e Riflessioni sopra le prime Considerationi di F. Stefano De Gl'Angeli Venetiano . . .* , Padoua, 1668.

[191] In the preface (*Al Lettore*) of the *Seconde Considerationi*, Stefano degli Angeli writes that though the claim of Manfredi to represent exactly the views of Riccioli is unbelievable, "nullo dimeno per dar questo piacere al non da me conosciuto, ne mai udito nominare Signor Manfredi, mostraro di crederlo." As we have seen Manfredi had published, in 1666, a booklet on the reform of the calendar. Thus Angeli *could* have heard about him.

[192] Stefano degli Angeli will attempt to show that Manfredi not only does not understand his, Angeli's, theory, but is even ignorant of the teaching of Riccioli and, as often as not, contradicts the latter.

[193] *Cf. Seconde Considerationi*, 71.

[194] Angeli has in mind the assertion of Manfredi that the motion compounded by the circular and the rectilinear ones (*cf. supra* p. 366) will not unite *perfectly* the properties of the two compounding motions—as this would be the case in purely geometrical kinematics—but only imperfectly. This means that, whereas in pure geometry the circular and the rectilinear motions, taken together, will determine a perfect curve, be it a parabola or a spiral, *in rerum natura* the trajectory of a body (real and not geometrical) moved by two different *impetus* will never be such a curve, at least not exactly but only approximately, because the two *impetus* cannot be blended together without weakening each other.

[195] *Ibid.*, 82.

[196] *Cf.* my paper quoted *supra*, n. 148 and Cornelis de Waard, quoted *supra*, n. 6.

[197] *Ibid.*, 82.

part; [198] then this first would not descend in order to unite once more with the latter. And likewise, that if one separated the larger from the smaller and transported the former into another place, but not outside the sphere of its activity, then the smaller piece would abandon its own place and would be drawn by the virtue of the larger.[199] He who would discourse in such a manner would perhaps not be far from the verisimilar, as this our Earth is, perhaps, *magnus magnet.*[200] And in the magnet we see that the larger draws to it the smaller placed inside of the sphere of its activity. And if, once united, the larger is transported into another place, yet still, inside the sphere of its activity, it will likewise draw the smaller to itself.

But Stefano degli Angeli is not a partisan of the hypothesis of attraction and he replies:

Although these are discourses of men of valor, nevertheless do not repeat them Sign. Ofreddi, if you do not wish to be laughed at. Besides, who assures us that in the magnetic motion [it is] the larger which draws the smaller to itself, and not much more this latter which moves itself in order to join the former,[201] because of some good which it receives from this union? Yet it is true that as this does not occur without motion and action it seems more reasonable to attribute the action of moving to the stronger part.

Ofreddi still insists and even improves his conception:[202]

But it could also be said that the action is common to both parts, that is, that the larger draws the smaller as much as this latter [draws] the former, but that finally the larger prevails, the smaller being unable to move the larger. Thus if to a string were attached a man and a child, the one at one end and the other at the other, though the child would pull the man, nevertheless, he would not move him from his own place, but would be pulled by the larger force.

Yet Angeli obviously does not believe in attraction [203] and answers: "Let us drop all these fantasies, which are of no weight, and let us pass to § 46," that is to Manfredi's criticism of Angeli's examples, criticism based on the distinction between the intrinsic and extrinsic principles of motion.[204]

CONT. He says in it that to join together two bodies really distinct moved by two principles, the one intrinsic and the other extrinsic, is beside the point. Wherefore the examples are, too, beside the point: [for example] that of running while carrying perpendicularly to the plane of the horizon a hollow rod in which a small ball would fall, and would be let fall. Or that of running while carrying a clepsydra with powder or water [in its upper part] which would fall in the container below the powder or the water.[205]

MATT. If they are beside the point, then it is the same concerning them and those that he brings himself.

CONT. He says, however, that they will be partly valid in so far as they show the diminishing of the descending speed.[206] Just as in the case of two equal bottles filled with water, which being allowed to run out of them, will flow out more rapidly from the bottle at rest than from the one which would be carried by a runner.

One would expect that Angeli would, contemptuously, reject Manfredi's assertion. Quite the contrary: he accepts it, or, at least, he accepts the fact alleged by him, giving to it, however, a somewhat different explanation: [207]

MATT. I do believe too, that it will so happen in these cases, because the horizontal motion is impressed anew on the bottle, and on the water, and first on this, then on that. Moreover, the water in its downward flow strikes the bottle which prevents it from coming down with the speed which is required by gravity.

Thus it is not the transverse motion as such that interferes with the descending one: it is its newness. Angeli thinks probably that the new motion, or *impetus* needs time in order to take hold of the body, and that during this time it can, indeed, hinder the other one. But not later on, when it is no more a disturbance, but, so to say, a regular feature of the case. Therefore

in [the case of] the motion of the Earth it would not be so; because the circular motion would be coeval with the Earth and with the heavy body; therefore, it would not impede the descending. And if it did impede it, the retardation would be proportional; that is, it would descend according to the squares of the times, but the spaces would be proportionally greater or smaller, according to their being compared to a greater or smaller circular motion.

This means that, even if the uniform motion of the Earth did slow down the descending one, it would affect neither the very fact of falling, nor its law; it would only affect the speed of descending. It would not, as Riccioli asserts and wants us to believe, prevent bodies from coming down; it would, at most, prevent them from coming down as quickly as they would if the Earth stood still.

To sum up, Riccioli's theory, according to Manfredi himself, is based on four pre-suppositions: [208]

---

[198] All the theories of magnetism and attraction, since Gilbert and Kepler and until Newton, assumed as something evident, that these forces have only a limited sphere of action; which, by the way, is in strict accordance with experience.

[199] The reasoning of Ofreddi implies the negation of the conception of natural places.

[200] That the Earth is a *magnus magnet* had been proclaimed by Gilbert and this theory had been highly praised by Galileo.

[201] The idea of attraction is a new one and appears only with Gilbert and Kepler; before them, attraction, as a force acting upon a body from outside and at a distance, had never been admitted and the magnetic action had been explained by a seemingly attracted body to join the seemingly attracting one. In the seventeenth century, just because it is an action at a distance, the conception of attraction was likewise rejected by the promoters of the new mechanical science.

[202] *Ibid.,* 83.

[203] Neither does Borelli, nor, as a matter of fact, Riccioli.

[204] *Ibid.* It is to be noted that the distinction between internal and external principles of motion is by no means denied by Angeli. What he denies is only the use made by Manfredi of this distinction.

[205] In these examples, the moving principles are *really* distinct, and therefore compatible, at least up to a certain point.

[206] Though compatible, they are not completely so; different motions hinder each other *as motions;* consequently, the velocity of the downward one will be diminished.

[207] *Ibid.*

[208] *Ibid.,* 99 sq.; *cf.* Manfredi, 88 sq.

The first pre-supposition is that, if the Earth turned around its center, then the terrestrial bodies, though separated from it, should equally turn around with it in the diurnal revolution.

This, objects Angeli, is certainly so according to the Copernicans; but it is not necessarily so. The earthly bodies, not attached to the Earth, could lag behind somewhat, as many people believe them to do, explaining thus the trade winds and the great oceanic currents. Besides, as Galileo long ago has shown in dealing with all these hoary and decrepit arguments against the Copernican system, the diurnal revolution, being mechanically common to all terrestrial bodies, is irrelevant to the processes occurring on the Earth.

The second supposition is, that even if the terrestrial bodies separated from the Earth turned around with equal speed with it, in order to conform themselves, as parts, to the motion of the whole, they would have to descend. Yet it is more probable that the *impetus* of the diurnal motion would prevail so much over gravity, or over the principle of descent, that it would completely hinder the action of this latter.

By no means, replies Angeli:

I am of the opinion that it would not hinder anything as long as the circular motion as well as the descending one were referred to the same center in all their parts. Nor would the circular motion be newly impressed on the heavy body, but [would be] uniformly perpetual.

But Manfredi has more arguments and analogies at his disposal. He says that:

If we should want to base our argumentation not upon fantasies, but upon [things] better known, we would see that a stone, turned around by a sling of two feet of length, and which in one beating of the pulse accomplishes a circuit of about 13 feet, becomes hindered by the velocity of the gyration and does not descend at all; and we would see the same in the case of a small vessel in which water is rapidly turned around on a string, because when it passes the superior semicircle of the gyration not a drop of water contained in it falls. Accordingly, the circular motion of the heavy body being so much more rapid, it should not descend etc.

Let us not misunderstand Manfredi; it is *not* the centrifugal force produced by the rotation of the Earth that, according to him would prevent the heavy bodies from descending—*in hypothesi terrae motae*—but only and solely the *speed* of the *linear* motion to which they would be subjected in this case. And it is not the weakness of this force that will be alleged by Angeli, but the physical difference between the two cases.

MATT. The first example, that of the sling, can be disregarded, because the heavy body is *quasi* imprisoned in it, and thus prevented from descending. The example of the vessel is better; but the case is so different from that of the motion of heavy bodies in the Copernican hypothesis, that it has nothing to do with it. *First*, the circular motion of the vessel is referred to a proper center, and gravity leads to that of the world. *Second*, the water in the vessel is, in part, contiguous to it and this contiguity hinders the separation, and the motion downward. *Third*, the air

hinders very much the separation of the parts of the liquid from each other, as we see it happen when we open the vessel at the bottom: the water is, to a certain extent, prevented by the air from descending and does not descend but in time.[209] Which is more certain to occur in the case of the very rapid circular motion. *Fourth*, because of the diversity of their position, the parts of the water endeavor to descend along different ways and must execute contrary motions which cannot produce their effects except in time, which cannot be admitted with regard to the velocity of the circular motion. I will explain this point more clearly. The circular motion is compounded by two semicircles, that is by the superior and by the interior. When the inferior semicircle is described, the parts of the water nearest to the bottom are the first, which, by their gravity, endeavor to descend, and the more remote gravitate upon them and press them. When hereafter the superior semicircle is described, the parts of the water more distant from the bottom are the first to descend. Thus if we consider well these inclinations to descend, we will see that in respect to the water they have contrary motions, although referred to the same center of the Earth; and that the act of descending of the inferior part is, with respect to the parts of the water, opposed to the act of descending of the superior part. But the passage from an act to another, opposite, cannot be made except in time, which is not the case when we are dealing with the velocity of the circular motion.[210]

CONT. The third supposition I believe to have the following meaning: the Copernicans suppose that the descending motion begins at once; and that whether the Earth moves or not a heavy body in one second of the hour descends 15 feet. It will not descend thus because for a certain time it will be supported by the diurnal impetus on the arc *BH*, as we have seen it happen, that the *impetus impressed on the ball of the arquebus or of the cannon, prevails so much over the gravity of the ball, that it does not begin to descend at once, but only after a long space*, and strikes the goal point-blank, proceeding for a long stretch on a horizontal line, and afterwards begins gradually to decline *from this very line* [describing a curve] *very similar to the parabola*.[211]

Manfredi, indeed, has learned nothing from the discussion of this problem by (and since) Galileo. But Angeli knows better:

MATT. Sign. Manfredi deludes himself if he believes that the ball is, even for a moment, thrown by the fire along a horizontal line. Having left the barrel of the cannon it begins immediately to descend, but at the moment of the departure the descending motion is very slow and the transverse very quick: wherefrom it results that, proceeding on a parabolical [line], which, in the beginning, because of its wideness, deviates very little from the

---

[209] Angeli intends to say that water does not start its descent immediately after—or, better, with—the opening of the vessel: it needs some time to overcome the resistance of the air.

[210] The parts of the water, according to Angeli are subjected by gravity to two opposite *impetus*, and the replacement of one *impetus* by another, contrary, one cannot occur in an instant, but needs time for its accomplishment. This is not the case in respect to the diurnal motion which is always there, and which is not changing, but uniform. There is no passage from an act, or a state, to its opposite. Therefore, gravity will act on the heavy body in the very instant of its release.

[211] The belief in the rectilinear motion of the bullet, which originated with the gunners is a good example of the disastrous effects of experiment upon theory. *Cf.* fig. 18, p. 364.

horizontal tangent, it strikes, as it is wont to say, point-blank. But separated from the cannon it will never move on the tangent prolonging the straightness of the cannon. As much would happen if the Earth moved. Never would the heavy body move only in a circle, if it were separated from its support, but would at once begin to descend.

The fourth supposition [212] is equivalent to Riccioli's famous argument from the force of percussion: on a moving Earth, the trajectory of the fall would be circular, the motion, therefore, uniform. But we are already acquainted with this argument. We do not need to deal with it once more, especially as Angeli has nothing to add to what he has said in his first *Considerationi*, namely, that the acceleration of *a body falling on a moving Earth* will be as real, and as notable, as on one standing still because in both cases the motion downwards will be along the selfsame radius of the earth. He adds, however, some well-chosen epithets and the statement that he believes to have finally and definitely proved the falsehood of Riccioli's circular theory of the fall and, thus, the worthlessness of all the conclusions based upon it. To treat him, therefore, as a crypto-Copernican is only to show as much stupidity as pride and bad faith. On the contrary, to uncover the paralogisms of Riccioli's reasonings is to serve the truth, and the true authority—that of Holy Scripture and the Holy Church—which has decided the question.

XI. BORELLI CONTRA S. DEGLI ANGELI

The story of the quarrel between F. Stefano degli Angeli and P. Giambattista Riccioli has somewhat distracted us from our main purpose: to trace the history of the determination of the trajectory of the falling body *in hypothesi terrae motae*. We have now to come back to our problem and to turn our attention, once more, to Borelli.

We have seen that, in the very first of the dialogues constituting his *Considerationi*, Stefano degli Angeli deliberately attacked Borelli's rejection of the spiral theory of fall, and that in the second one he added to his criticism a particularly nasty remark. Small wonder that Borelli replied immediately, and not very amiably, by an open letter addressed to Michelangelo Ricci [213] as

the man to whom, even before the publication of his book on the Force of Percussion, he had communicated his views on the problem.

Having stated that the criticism of Stefano degli Angeli obliges him to answer, Borelli plunges immediately *in medias res*: [214]

He [Stefano degli Angeli] begins at page 29 of his dialogue to wonder that I have denied that the above-mentioned mixed motion could be performed along a particular spiral which some moderns have treated. These, be it known to Your Grace, are Mr. Fermat as P. Mersenne reports it on page 5 of his *Phenomena Ballistica*, [215] the demonstration of which, adds this latter, has been sent to Galileo, and it is possible, that copies also were sent by him to his friends, as was his habit; he [Mersenne] says also that the space of this spiral to the sector of the circle which comprises it has the proportion of 8 to 15, just as it is advanced by P. Angeli. In order to show, then, that this line is really a spiral, the said Father [Angeli] supposes that the moving body is carried by the motion of the semi-diameter, which [motion], he says, is in the moving body by participation, and thus the moving body, in truth, is moved transversely not by a uniform motion, but [by a motion] successively retarded, and this, he says, is verified by the fall of heavy bodies; this proposition he assumes [216] simply as being true, there being no other reason for confirming it but his own and sole authority.

I, on the contrary, believe that it is impossible that the transverse motion conferred on the stone by the supposed revolution of the summit of the tower, or of the mast of the ship, around the center of the Earth, can proceed in slowing down in the proportion in which it approaches more and more the terrestrial center, where finally it has to be extinguished; but [I] believe [that] in whatever place of the descent it finds itself, it must conserve the same degree of transverse velocity, and consequently traverse equal spaces in equal times on all the circles that it traverses. It seems to me that this is confirmed by numerous experiences and by the *right* reason; because I see that whatever *impetus* and degree of velocity be conferred on a moving body, though it can indeed be weakened, and extinguished either by a contrary *impetus* or by some resistance it encounters, yet not merely by the fact that it changes its direction. Thus it is seen that a moving body which moves with its degree of acquired velocity, either on a straight line, or on a circumference of a circle, when it occurs that it simply changes its path, either by reflection, or because it curves its path inward more than at first, and moves on smaller circles, conserves the same velocity, which it had at first. Thus a ship which has acquired a determined rate of speed either from the wind or from the oars, when it turns and describes more curved paths, is seen to run with the same *impetus* as at the beginning, and the same is observed in the cases of flying birds, and of all the projectiles; and if somebody wants an experience similar enough to that which we are discussing, he could take a pendulum, such as *ABC* [fig. 20], make its thread pass through the ring *B*, attached to the ceiling of a room, and then turn this pendulum around impressing on it a determined rate of speed in such a manner that the lead ball would describe the circle *ADE*. If then, with his hand, he

---

[212] *Ibid.*, 104; Manfredi, 92.

[213] *Risposta di Gio: Alfonso Borelli Messinese . . . Matematico dello Studio di Pisa Alle Considerazioni fatte sopra alcuni luoghi del suo Libro della Forza della Percossa del R.P.F. Stefano degl' Angeli Matematico nello Studio di Padova, all'Illustrissimo, e Dottissimo Sig. Michel Angelo Ricci*, Messina 29. 11. 1668, p. 1: "Credo che V. S. Illustrissima avérà molto primo da me veduti certi Dialoghi dal Dotissimo Patre Stefano de gl'Angeli, scritti di certa dimostratione contra il sistema Copernicano, et in detto libro si è compiaciuto di considerar quella digressioncella, che io fa alla facia 108 del mio libro della forza della percossa, doue io considero il moto misto del transversale circolare equabile, e del perpendicolare verso il centro del cerchio uniformamente accelerato, dal qual moto misto mi recordo hauerne scritto à V. S. Illustrissimo da Pisa, primo che il mio libro si stampasse."

[214] *Ibid.*, 2 sq. The style of Borelli is atrocious. He is probably the worst writer of the seventeenth century, not excepting even Cavalieri.

[215] Obviously a misprint—5 for 50.

[216] *. . . assume egli solamente per cosa vera*. Borelli wants to say that Angeli accepts this proposition without proof, and that it is only assumed as true, without being so.

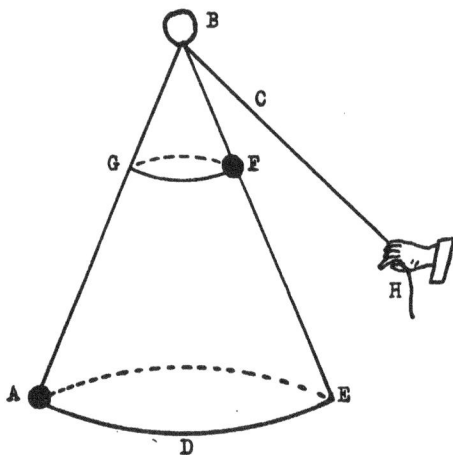

FIG. 20

ABC (fig. 21); and let it be fixed on the axis *CD*; let this axis be passed through a hole *E* of the plane *FG*; and let its end *D* be placed on a socket in the floor, in such a manner that this vase be able to move (to revolve) around its vertical axis; let this vase, in its inferior part, have a circular ridge, such that it would be able to hold a small ball, made of wood, or of another matter, and allow it to roll around in it; let us then put this ball *A* on the superior rim of the vase in *A*. Let us then turn this vase around until it acquires a certain, determinate, velocity. There is no doubt that the ball *A* will acquire the same velocity as the rim of the vase, and that, therefore, it will remain on the same place on the rim. Let us now lower the ball *A* to the point *H*. Then, continues Borelli, if the velocity of the vase is maintained, we shall see that the small ball *A* will not remain in the same place, but will rush forward to *O*; that is it will out-run the vase by the arc *HIO*, such that this arc *HIO* together with the circle *HI* will

would draw the end of this thread *C* in such a manner that the length of the pendulum *BF* be the fourth part of *AB*, then he will see the ball describe the circle *FG* in a shorter time, that is in the half of that which it needs to complete the circle *ADE*, and yet the velocity on *FG* will be the same which the ball has had in *A*, and on the contrary, lengthening the pendulum from *G* to *A* we shall see it starting at once to extend the revolutions of the ball, but always in obedience to the law that the speed be the same in every place, neglecting indeed the variation depending on the approach to the center of the circle *AD* made by the ball, which, in our case, is of no importance.

To the example of the circular pendulum Borelli adds that of a ball running on different levels of an inverted cone, and turning around so much the faster as the level upon which it moves is lower, and yet moving the whole time with the same linear, or circular (still considered as identical), velocity, which in this way becomes clearly distinguishable from the angular one.[217]

The following observations—explains Borelli [217a]—will be no less clear and evident. Let us take a conical vase

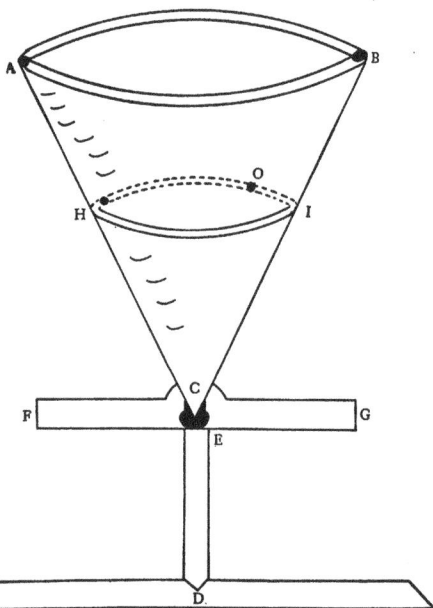

FIG. 21

---

[217] *Ibid.*, 4 sq.

[217a] *Ibid.*, 4 sq.: "Non meno chiara, e evidenti sarà questa esperienze; prendasi un vaso conico à guisa di bicchiere *ABC*, il quale sia saldato nell'asse *CD*, e questo sia infilzato nel forame trasversale *E* d'una stanza *FG*, e con le vertice inferiore *D* si appoggi in una forame del pavimento in maniera, che tutto le vaso sia volubile intorno al suo asse perpendicolare all'orizonte; abbia poi le detto vaso nella sua parte interna una zona ò risalto circulare *HI* nel qual possa sostenersi, e girare una pallottolina di legno ò d'altra materia; posta poi la detta palla *A* nell'orlo supremo del vaso in *A* cominci a rivoltarsi insieme col vaso sin che arrivi ad una determinata velocità, non ha dubbio che la palla *A* acquisterà la medema velocità, che hà l'orlo supremo del uaso *AB*, cioè si manterra nel medemo segno, ò termine del lato del cono *AHC*, ora se in questo stato si lascerà precipitar la pallottolina *A* sino ad *H*, e se manterra la velocità del uaso nel medemo grado di prima, si vedra, chê la pallottolina arriuata in *H* non si fermara nel medemo lato del cono *AC*, ma scorrerà auanti fino in *O* in maniera, che l'intero cerchio *HI*, insieme con l'arco *HIO* sia eguale, e tutta

la circonferenza *AB*; segno evidente, che il grado di uelocita, che aveva la pallottolina in *A*, conseruandosi anco in *H*, è necessario, che trascorra spazio equale a quel primo nel medemo tempo d'un intera revolucione del uaso, si che, non perche si conduce più al basso alla circonferenza d'un cerchio minore per questo perde punto di quella velocità che la uertigine del uaso sia egualmente ueloce, tanto quando la palla è nel orlo *A*, quanto dopo esser caduta in *H*, si potrebbe porre un'altra pallottolina in *B*, la quale, se dopo esser caduta la palla *A* in *H*, perseuera l'altro nel termine *B* nel lato *CB* senza scorrere auanti, ne rimaner in dietro sarà segno indubitato, che la vertigine del uaso e uniformemente ueloce, e simile à se medesima."

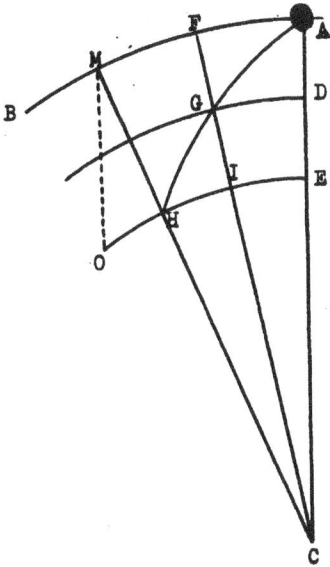

Fig. 22

be equal to the circumference AB. This shows clearly that the ball A conserves at the point H the velocity that it had at A, that, therefore, it must, necessarily, traverse in the same time of a complete revolution of the vase a space equal to that which it traversed before it was lowered to H, and that it does not lose speed because it has to move on a smaller circumference.

Having achieved his purpose, Borelli goes on to the discussion of the concrete case of a stone falling from the top of a tower or the mast of a ship: [218]

Coming now to our case, let C [fig. 22] be the center of the Earth, the circumference of the Equator be EH, and let AE be the height of the tower, or else [that] of the mast of the ship, and let it be admitted that AE, together with the semidiameter EC, perform a uniform circular motion along the circumference EH; let us now drop a stone from the summit A. We agree, myself and the R. P. Angeli, that such a stone leaves the summit of the tower A with two *impetus*, the one communicated to it by the preceding circumgyration of the semidiameter CEA, and this one is uniform, that is apt to traverse equal spaces in equal times; the other *impetus* is that which it has as a heavy body apt to move itself to the center of the Earth C with a uniformly accelerated motion. Thus far we are both in agreement. The only thing which is controverted, is [the question] of whether the motion, transverse in respect to [that] of approaching toward the center of the Earth performed by the stone, will be successively retarded in proportion to the distances to the center, that is, supposing that the transverse *impetus* were apt to traverse in one second of time the arc AF at the summit of the tower, whether this heavy body brought lower down, for instance,

to G, would in as much time traverse there an arc smaller than FA and equal to GD; and whether, transferred to H, it would pass in the same time an arc equal to EI and so on, until, led to the center C, the said transverse *impetus* will be completely extinguished, and reduced to rest. And in truth, I understand very well, that if the line AE were a hollow rod, containing the falling leaden ball, then the same rod circumgyrating together with the semidiameter CA would transport the said ball of lead wherever the rod is moved, and thus, when AE arrived in FI, the ball A would have descended toward the center through all the space FG, and, in virtue of the transverse motion it would have traversed the arc DG, smaller than the arc AF, and in the second time, the semidiameter together with its rod being brought to the place MH, and the ball having arrived in H, [it] would, in virtue of the transverse motion, have traversed the arc IH equal to IE and smaller than DG and so on. All that would follow by necessity because the ball is forced, willy-nilly, to move in the length of the channel AE in which it is imprisoned, and thus it would be true that the transverse motion of the ball would be really the same as that of the semidiameter CA, and would be in the ball only by participation, so that the said ball would not possess of itself a determinate rate of velocity, but would have it by participation, possessing continually as much [velocity] as the semidiameter CA, to the motion of which the ball is forced to obey, would have had communicated to it. And I have no doubt that in this case the mixed motion of this ball would describe the spiral of which Mr. Fermat and P. Angeli are speaking, and this would likewise follow if, in purely geometrical reasoning, it were assumed that the point A moves simply on the semidiameter AC with a uniformly accelerated motion, while the semidiameter AC moves on the circumference of the circle with a uniform one. [219]

But the case we are dealing with is quite different, explains Borelli. We are not studying, in purely geometrical manner, the motion of a point on its rotating radius—a very interesting problem, to which Fermat has given a brilliant solution—nor are we studying the motion of a ball running inside a hollow rod or otherwise mechanically attached to a physical body espousing the direction of the radius-vector, and thus obliged to follow its motion. We are dealing with a heavy body which is not attached in any way to the Tower or the mast it is falling from. In our case the radius vector is a purely imaginary line which in no way affects the motion of the body. This body finds itself, at the beginning of its fall, endowed with a certain determined transverse *impetus*, and as the air, moving with the Earth, does not oppose any resistance to its motion, this *impetus*, indelibly impressed on the body, will remain identical to itself. Therefore, the body will not slow down but forever continue its motion with the same speed, that is, traverse the same distances in the same amount of time, in spite of the fact that it will move on ever diminishing circumferences. It is, in this case,

---

[218] *Ibid.*, 6 sq. It is interesting to note the difference between the drawing which Borelli gives *now*, and that which he gave in his *De Vi Percussionis, cf. supra,* p. 358.

[219] It is interesting to note the exact similarity of Borelli's reasoning—and example—with that of Manfredi. And yet, it is hardly possible that the latter's book, of which the preface is dated January 10, 1668, could have been used by Borelli who dates his own work February 29, 1668. Besides, Borelli says in the Appendix to his letter (p. 24) that he has just received Manfredi's book, that is, after the completion of his own work.

perfectly ridiculous to think that it will slow down its motion because it approaches the center. To think so is only to demonstrate one's inability to understand the difference between pure and applied (mixed) mathematical sciences, such as mechanics, acoustics, and so on, an inability all the more damning as the problems of the relations of these sciences have been clearly explained in the very work which has been criticized by Angeli.[220] For the true reasons and careful experiments upon which dynamics is based in this book, Angeli substitutes his own, unsupported opinion, his "I think, I believe, So it seems to me," etc.

People ignorant of dynamics, who cannot even understand that, in substituting for free fall either a mechanically determined movement, or the abstract motion of a point on a geometrical line, who change the initial data of their problem and do not abide by their initial hypotheses, should keep quiet. If they do not, they are bound to say a lot of nonsense, as it is the case with the Reverend Father Stefano degli Angeli.

Having dealt with his opponent in this, rather rough, manner—Borelli, as is well known, was always a man "nice to deal with," and, when a member of the Accademia del Cimento, managed to quarrel with everybody, even with Toricelli—Borelli goes over to the discussion of the last objection raised against him by Angeli, namely, that, if his theory were true, the falling stone would not fall at the foot of the Tower but outrun it and deviate to the east.[221]

I am now coming to an opinion which seems so absurd that only people totally deprived of human sight and understanding could maintain it. I said[222] that the stone falling from the top of the tower, or else from the top of the mast will gyrate on a great circle around the center of the Earth (I am always assuming this as a mere hypothesis)[223] and will not accomplish its fall along a line designed by this same tower perpendicularly to the terrestrial surface, but must deviate from it, so that the apparent path of such a descent will not run directly to the center, but will decline from the terrestrial semidiameter, to which it was united at the beginning. And in truth, I do not deny that this has troubled me so much that, after I finished the printing of my book, I was moved to modify the hypothesis as well as the line described by the above-mentioned heavy body;[224] yet it cannot be denied that the common concept[225] does not depend on a simple imagination, or on a prejudice, and that one cannot reject it [because it leads to the assertion of a deviation of the falling body from the perpendicular] without having taken the trouble to consider attentively what and how great would be the deviation from the

perpendicular of the falling stone to the surface of the Earth. If we do it we will recognize clearly that the apparent deviation from the perpendicular, wherever there would be one, would be absolutely unobservable because of its smallness. In order to understand it clearly, we shall use the same figure as before. The tower AE is 240 feet high, and it is supposed to circumgyrate with the terrestrial semidiameter CE on the arc EH of one minute of the Equator in such a manner that the tower moves to the site HM and that in this time a ball of chalk[226] falling from the top A arrives at the Earth with two motions, i.e. with the transverse motion, of which the uniform *impetus* is measured by the arc AM and with the descending *impetus* on the perpendicular AE. I say that in this case the ball will not fall precisely upon the lowest place H of the perpendicular to the horizon HM traced on the face of the tower, but that, falling along the line MO, it will outrun it somewhat, [and that] the arc EO will have to be equal to AM and thus greater than EH. I must show now that the deviation HO, on account of its smallness, cannot be observed, because as the terrestrial semidiameter CA is supposed to be of 23,367,468 Roman feet (antique) and the tower EA of having a height of 240 of the same feet, the proportion of the terrestrial arc EH of a minute of the Equator to the arc AM traversed by the top of the tower, will be the same as that of CE to AC; now this transit is made in 4″ of the hour: therefore, admitting the arc HE to be of 6,797⅓ Roman feet, the arc AM, or thus EO [will have] 6,797⁴⁸⁄₁₂₀ feet and consequently the excess HO will be ⁹⁄₁₂₀ foot, which is ⁹⁄₁₂ inch, and thus less than an inch.

Now, explains Borelli, this is too small a deviation to be observed, because the margin of experimental errors that are bound to occur is much greater even if for the dropping of a stone held by the hand—which cannot be perfectly at rest in respect to the Tower or the mast—we substitute the fall of a ball attached to a thread and placed inside a pyramid completely enclosing it (this in order to prevent the disturbing effects of the turbulent motion of the air), the fall being occasioned by the cutting of the thread.[227] There will still be so very many perturbing factors that the deviation will remain experimentally unobservable, and the falling body will be perceived by us as moving along, or parallel to, the Tower or the mast.

But there is something else, and even something much more important to be considered, namely, the fact that the transverse impetus of the falling body will not be *circular*, as it has been admitted, but *rectilinear*:[228]

It is necessary to remind ourselves of a certain natural property of the circular motion which has the faculty of extruding and of removing the movables from the circumference on which they start moving, [and to drive them away] on a straight line tangent to the circle in the point of separation, every time it happens that the moving body detaches itself from this circumference. Thus we see in the case of a stone carried by the revolution of a wheel, that every time it detaches itself from this latter, the preceding

[220] Borelli means, of course, his own book: *De Vi Percussionis*.

[221] *Ibid.*, 12 sq.

[222] As a matter of fact, he did not; Angeli deduced it from his premises.

[223] Borelli has been strongly—and rightly—suspected of Copernicanism.

[224] He did not modify his hypothesis; but he did modify "the line," at least in the drawing.

[225] *Il commun concetto*, i.e., not the commonly held conception, or the concept of the common sense, but the axiomatic concept of reason.

[226] Allusion to the experiments of Riccioli for which he utilized the Torre degli Asinelli in Bologna.

[227] Borelli has a fine sense of the conditions required for a good experiment. His technique—he learned it probably at the Academia del Cimento—is just as good as that of Huygens or Hooke.

[228] *Ibid.*, 16 sq. *Cf.* fig. 23.

circular *impetus* degenerates [229] into a rectilinear one, though without declining, that is without making any angle with the circumference, on which it moved at first. Now this being commonly received as something certain and evident, when I consider the top of the tower or the mast of the ship *A* revolving around the center of the Earth *C* with the semidiameter *CA* on the circumference of the great circle *DA*, I understand that the stone or the ball placed on the top *A*, does in virtue of the circumgyration *DA* possess the same *impetus* and velocity which the top of the tower *A* has, and if it happens that the said ball detaches itself from the finger which supported it in this place, that is, if it is abandoned in a fluid medium,[230] it seems necessary that this ball should continue the impetus acquired by the circumgyration *DA* no longer on the circumference of the circle *AB*, but on the straight line *AH* tangent to the circle in the point *A*.[231]

The results of this very important correction of the initial analysis of the dynamics of fall is to enable us to see that the real deviation of the falling body from the perpendicular will be even much smaller than we have just calculated, so small as to be practically not only unobservable, but nonexistent.

Indeed, let us reconsider the case. If the heavy body released from the top of the tower were deprived of gravity (we can easily imagine it),[232] nobody would doubt that it would continue its motion, not on the corresponding circle, but on the tangent to its point of release, just as it happens when a body turned around by a rotating wheel detaches itself from it: it flies away along the tangent. But the real case of the heavy body is by no means different. It has to be considered as endowed with two *impetus,* one along the tangent, and the other toward the center of the Earth, and as these *impetus* or corresponding motions, the uniform one along the tangent, and the accelerated one along the radius, do not interfere with each other, the body will move on a curve compounded by both.[233] To determine this curve we have to modify our drawing and to trace, besides the circumference *AB* (the path of the summit of the Tower, fig. 23), the tangent *AH* and the secants *CE, CF, CG* and *CH,* passing through the points *E, F, G,* and *H,* corresponding to equal distances *AE, EF, FG* and *GH* which would be traversed by the body (each in 1″ of the hour) if the *impetus* of gravity did not move it toward the center *C.* Then, in order to find the position of the falling body at the end of each second, we have only to determine the points *M, N,*

Fɪɢ. 23

O, P, distant, respectively, from the (imaginary) points *E, F, G, H,* by 15, 60, 135, and 240 feet. Accordingly,[234]

. . . the true motion of the ball, compounded by the transverse and the descending, will be made on a curve drawn through the points *A, M, N, O, P* and thus in every place the falling ball will move skirting the same semidiameter, and consequently the same line traced on the face of the tower; for in the same time of 1″ in which the transverse *impetus* pushes the stone on the tangent *AE,* the *impetus* of the gravity transports it on the secant *EC* from *E* to *M* and this secant *EC* forms with the tangent *EA* an acute angle because the angle *EAC* is a straight one in the same triangle.[235]

It results therefrom that the descending *impetus EM* pushes, so to say, the body back to *A.* The point *M* corresponds to a rectilinear motion to the point *R* and to a descent, on the perpendicular line *RM,* equal to *AS.* After the second time, the point *N* will, accordingly, correspond to a (rectilinear) motion to *T* and a perpendicular descent *TN.*[236]

Thus we shall be able to say without a sensible error that the ball and the direction of the tower, or of the mast of the ship, will be placed on the same secant *ES,* and it will be so for all the other secants *FC, GC, HC*; and, therefore, on the whole transit along the curve *AP* the falling ball will always find itself placed so as to skirt the same

---

[229] The expression is rather characteristic!

[230] In the air, in which it moves freely.

[231] *Ibid.,* 19 sq.; 22 sq. It follows from Borelli's theory that even on the Earth, and not only in the skies, the motion of the heavy body will be compounded by a motion to the center and a motion along the tangent. That is exactly what Robert Hooke will tell Newton somewhat later; *cf.* my paper quoted, n. 1.

[232] This is by no means easy for a seventeenth-century mind. A body deprived of gravity is no more a real body and cannot, therefore, receive an *impetus; cf. infra,* p. 381.

[233] On the assimilation of gravity to any other force by the pupils of Galileo *cf.* my *Etudes Galiléennes* 3, *App. L'élimination de la pesanteur.*

[234] *Risposte,* 18 sq.

[235] The angle formed by the intersection of the secants with the straight line *AH* becomes more and more acute, and therefore the *impetus* along the secants toward *C* draws the moving body more and more toward *A,* i.e., toward the tower.

[236] *Ibid.,* 19 sq.

tower, or mast of the ship, along the same straight line perpendicular to the horizon. Besides, as in an arc that does not exceed 1″, the secants are not notably greater than the semidiameter, we can take the EM as equal to IM as I will show hereafter. Thus, though in the first [period of] time of 1″ of an hour the ball will have traversed the space IM of 15 feet from the top of the tower, and likewise at the end of the subsequent 1″ of an hour the ball descending from the top of the tower will have traversed the space KN of 60 feet, precisely equal to FN, and though we will say the same about the subsequent spaces, still it will be averred that, in the case of the gyration of the semidiameter AC, the ball falling from A will never deviate from this semidiameter. This [conclusion] follows from two suppositions [which are] not arbitrary as are those that the geometers are accustomed to make, but are in perfect accord with the laws of Nature, because it is a natural thing, as it has been said, that the circular gyration confers on a moving body (after its separation from the circumference) a rectilinear direction by the tangent as it has been said. Moreover it is in conformity with the habit of Nature that velocity acquired by a preceding motion can persist [in the moved body] only in making its progress with a uniformly rapid and equable [motion]; that is, that in equal times it traverses equal spaces in whichever place it finds itself as long as it does not encounter any cause which slows down or hinders its motion.

Borelli is well aware that, from the strictly mathematical point of view, the proposition deduced by him is not quite correct, and that, if the fall of the heavy body should be continued to the center of the Earth,[287] it would deviate considerably from the radius vector, at least in the middle part of the descent. But in our case, that is in the case accessible to experience, the deviation will be as nonexistent. Indeed, for an arc of 1″ it will be less than $\frac{1}{250}$ inch.

We must, however, take into account the fact that we have put the starting point of the downward motion in the points E, F, G, H, situated on the tangent and not in the points I, K, L, B representing the corresponding positions on the Tower. It seems to follow therefrom that, in order to arrive at the foot of the Tower, the falling body would have *more space* to traverse, and therefore would need *more time* for doing it, and thus strike the ground later if the Earth moved than if it stood still.[288] Yet it is easy to see that this excess of the secant on the tangent, even when corresponding to . an angular displacement of the Tower of 1″ performed in 4″ of time, will be not larger than 11 inches. This excess, compared to the whole distance of the descent—240 feet—is so small that the time difference, something about 27½‴‴, will be perfectly imperceptible and

we will not be able to measure it even with the quickest pendulum.[289]

Can it be supposed that a man will be able to observe exactly with a small enough pendulum, and be certain not to make errors in the measure of so imperceptible a time? Those who are experienced in the measuring of time made in this way know very well how easy it is to make an error of some third minutes, not to speak of fourth minutes which even the imagination does not reach. Now because this difficulty of the measuring is unavoidable, we shall be bound to confess that in the measures of the times of all the falls IM, KN, LO, BP, there will be committed some small errors, during which can very well be traversed the excesses of the said secants which, as it has been said, must be traversed in absolutely imperceptible times. And thus, being obliged to rely upon the judgment of our senses, we cannot but confess that in this hypothesis, in which the transverse uniform motion is made by the tangent of the circle AH, and the falls are made on secants of an arc as small as one minute, it is very possible to save the oblique path AP of the falling ball, with all the circumstances, that are shown by experience, that is, to maintain that the falling ball will always be skirting the face of the tower, and that the spaces traversed in the falls will be measured by a duplicate proportion of that of the times.

XII. ZERILLI CONTRA S. DEGLI ANGELI

If Borelli imagined that his booklet, so curiously thought out and so badly written, would convince, or at least silence, F. Stefano, and thus bring the polemic to an end, he was unduly optimistic and misjudged completely the character of his opponent.

Stefano degli Angeli instead of keeping quiet—something which he, indeed, could not do, because he had been practically accused of plagiarism by Borelli—replied by publishing his *Terze Considerationi.*[240] Yet, if in so doing he, in his turn, hoped to have the last word in this controversy with Borelli—as he believed he had had in that with Riccioli—he was very much mistaken. Borelli, indeed, let the matter drop, but a pupil of his, a certain Diego Zerilli,[241] descended into the arena in order to defend the position and theories of his master. And Riccioli, encouraged probably by the development of the strife, joined in the fight, under his own name this time, by publishing, in Latin, a dignified

---

[287] Borelli, in spite of the substitution of the movement on the tangent to that on the circumference, believes, nevertheless, that the heavy body will reach the center of the Earth.

[288] It is rather strange that Borelli does not notice that all these considerations are practically meaningless, as (a) it is impossible to compare experimentally the two cases, i.e., set the Earth in motion if it is at rest, or bring it to rest if it moves, and (b) the measurements of the speed of the descent are made on *this* Earth.

[289] *Ibid.*, 22 sq.

[240] *Terze Considerationi sopra una lettera del . . . Signor Gio: Alfonso Borelli, scritta in replica di alcuni dottrine incidentemente tocche da Fra S. de gl'Angeli . . . nelle sue prime considerationi sopra la forza di certo argomento contro il moto diurno della terra,* Venezia, 1668; cf. n. 130. The *Terze* and *Quarte Considerationi* of S. degli Angeli, as well as the writings of Zerilli and Riccioli, are exceedingly rare. I shall quote them, therefore, in the original.

[241] Practically nothing is known about Diego Zerilli. According to Gio. Cinelli, *Bibliotheca Volante* 4 : 386, Calvoli, 1747, he became in later years professor at the University of Pisa : "Era assai giovane il Signor Dottor Zerilli quinde diede in luce questo suo dottissimo opuscolo. Ora è un de' maggiori ornamenti dello studio Pisana, ed è con ragione amato, e stimato da chi che sia, toltone alcuni pochi, che non conoscono, o fingono di conoscere il merito di esso."

and rather pompous, *Apologia*.[242] Thus poor Father Stefano found himself in a worse situation than ever, having either to accept the defeat, or to continue the struggle defending himself against these new—and old—opponents. This he did, in the two *Dialogues*—the sixth and the seventh—of his *Quarte Considerationi*.[243]

In my analysis of these resurrected polemics I shall follow the chronological order. Thus, I shall first give an exposition of Stefano degli Angeli's answer to Borelli, together with the counter attack of Zerilli, then pass to the restatement of Angeli's own views as against those of his opponents, and conclude with Riccioli's *Apologia* and Stefano degli Angeli's final reply.

As a matter of fact neither Angeli's *Terze Considerazioni*, nor Zerilli's *Confermazione*[244] presents many new considerations and arguments concerning the problem of the trajectory of the fall; yet they bring *some*, and they are very interesting because they present to us the ultimate stage—before Hooke and Newton—of the discussion.

Angeli starts by explaining why, having decided to write no more on this subject after the publication of his *Seconde Considerationi*, he does not abide by his decision. The reason is that, Borelli having in his *Risposta* cast some doubts on Angeli's ingenuity and even upon his honesty, he has to reply in order to prove to the world his perfect sincerity and good faith, especially as his criticism of Borelli—whom he deeply admires—was not unprovoked. It was Borelli, indeed, who, in a letter to Michel Angelo Ricci, long before the publication of the first *Considerationi*, had attacked the theory developed there by Angeli.[245]

Father Stefano is particularly annoyed by Borelli's insinuation that he did not invent his theory, but borrowed it from Fermat *via* Mersenne.[246] He answers that the story told by Borelli about Fermat's solution of the problem, its publication by Mersenne, and its communication to Galileo, is completely "new" to him, as he does not possess the work of Mersenne and has seen it only in a cursory manner. But he is happy to be in agreement with so famous a man as Fermat, of whom, indeed, he has never seen anything, but whom he always has heard praised as a man of quite outstanding abilities.[247]

Angeli continues by protesting against Borelli's accusation of having based his theory of the diminishing transverse velocity of the falling body solely upon his own authority, and not on proof.[248] It's just the opposite which is true. The opinion held by him is based on the authority of Galileo, and it is common to all those who have defended the Copernican theory against the objections raised by its critics,[249] whereas that of Borelli is strictly his own.[250] To which Zerilli replies, not without reason, that this is beside the point: the problem being not whether this opinion is held by those who defend Copernicus, but whether it is true.[251]

Irked by Borelli's reproach of not understanding the difference between "pure" and "mixed" (applied) mathematical sciences, Angeli suggests a physical cause of the retardation, namely the resistance of the air to the transverse motion of the falling body, which must increase in proportion to the body's approach to the center of the Earth. In this approach, indeed, it penetrates into regions of the air that move ever slower—the air moves with the Earth, as a whole, and has everywhere the transverse velocity of the corresponding point of the radius-vector—and therefore it resists more and more, the relatively to it, ever quicker motion of the body in question.[252] Nobody, indeed, doubts the resistance of the medium, that is of the air, to the progressive motion. Even Borelli, though he believes that a moving body conserves its speed in spite of changing its direction, recognizes, and even teaches, that in the air

---

[242] *Apologia R. P. Io: Bapt. Ricciolii, Societatis IESU Pro Argumento Physicomathematico contra Systema Copernicanum*, Venetiis, 1669, *cf.* n. 130.

[243] *Quarte Considerationi sopra la confermatione d'una sentenza dal Sig. Gio. Alfonso Borelli . . . prodotta da Diego Zerilli . . . e sopra l'Apologia del M. R. P. Gio. Battista Riccioli . . . espresse dal medesimo S. degli Angeli, Venetiano.* . . . In Padoua, 1669; *cf.* n. 130.

[244] *Confermazione d'una sentenza del Signor Gio: Alfonso Borelli M. Matematico dello Studio di Pisa di nuouo contradetta dal M.R.P. Fra Stefano De Gl'Angeli Matematico dello Studio di Padoua nelle sue terze considerazioni prodotta da Diego Zerilli*. In Napoli, 1668; *cf.* n. 130.

[245] *Terze Considerationi*, preface, "*Al cortese lettore*. Havevo constantemente determinato di non scriver più altro in simile proposito, e hauerei esequito quando il Signor Borelli hauesse replicato cose tali, che tù le potesse giudicare dalla sola lettura, ma hauendo riempiuta questa lettera di questi sospetti della mia ingenuità, e modo di procedere, non posso, non deuo tacere in conto alcuno se non voglio dar a vedere d'acconsentire, essendo tropo geloso, che il mondo tutto sapia, che l'hò sempre riuerito, e stimato quanto meritano le sue gran virtu."

[246] *Cf. supra*, p. 371.

[247] *Terze Considerationi*, p. 3: "Havendola adunque trascorsa hò notato nel principio d'esso vna cosa, che mi è riuscita nuova, cioè che il P. Mersenno alla faccia 5 de suoi Fenomeni Balistici dica, che il Signor di Fermat habbia considerata la linea, che descriuerebbe il graue cadente, supposto falsamente muouesi la Terra, e che gia molti anni ne sia stata mandata la demostrazione al Galileo. Questa cosa mi è riuscita, come hò già detto, nuoua perche ne io tengo l'opere del P. Mersenno, ne mai l'ho vedute che alla sfuggita. Hò però gran piacere d'essermi accompagnato in questa speculatione con huomo tanto famoso, qual è stato Sig. di Fermat; di cui se bene non hò mai veduto cosa alcuna, hò però, e gia molti anni quando mi ritrouauo in Roma, dall illustrissimo, e dottisimo Signor Michel Angelo Ricci, e poi successivamente da molti altri peritissimi Geometri, vdito celebrarlo per soggeto realmente singolarissimo."

[248] *Cf. Terze Considerationi*, 5 sq.: Borelli says (*cf. supra* p. 371): "che confermi questa propositione con la sua mera autorita, mentre credo, che sia singolare."

[249] *Ibid.*, 6: "Il Galileo, Gassendo e tanti altri giudicano che il vento perpetuo verso Occidente nella Zone torrida prouenga dall'aria, che separata dalla Terra non seguiti totalmente il suo moto verso Oriente."

[250] *Ibid.*, 6: "Li Copernicaei sudano la fronte à conservare la cossistente velocità, . . . ma niuno, certamente, che io habbio veduto concede maggior celerità, come vuole il Sign. Borelli."

[251] *Cf. Confermazione*, 8.

[252] *Cf. Terze Considerationi*, 8 sq.

the *impetus* of the moving body diminishes continuously since the very first moment of its motion and, finally, is completely extinguished.[253]

The reasoning is very subtle, replies Zerilli. But will this retardation be sufficient in order to slow down the motion of the falling body in the proportion needed by Angeli? Obviously not, because the resistance of the air, as shown by the experiment of the pendulum, is much too small, and because, in any case, it will not be able to retard the motion of the said body so as to keep it on the radius-vector. Besides, Angeli seems to recognize it himself: he deems it necessary to allege a second cause of retardation.[254]

This second cause, which is by far the more important, is simply the necessity, for the heavy body, to remain on the same straight line (radius) in its motion towards the center of the Earth. This, says Angeli, because, as a matter of fact, the falling stone, when it separates itself from the Tower, has not simply a transverse circular velocity or *impetus*, but a relative velocity toward the center, which must continuously proceed in the same way and maintain it on the same semidiameter on which it started its motion.[255] If one should want a supplementary proof of the validity of this assertion one could make an appeal to Kepler's theory of magnetic attraction: isn't it clear that these forces of attraction which act between the heavy body and the Earth or between the magnet and the piece of iron will always draw them towards the center, and thus maintain them on the same perpendicular?[256]

But, answers Zerilli, is that proof? Is it anything else than a restatement of Angeli's assertion that the falling stone, while turning round, has always a velocity which maintains it on the semidiameter? The Reverend Father is only repeating the same things in different words and pretending he has given a demonstration

whereas he has merely reformulated his own opinion without backing it by any reason whatever.[257]

Indeed, insists Zerilli, Father Angeli is probably himself aware of the weakness of his arguments. He is recurring, therefore, "to I don't know what theory of Kepler"[258] and saying that the phenomena produced by the magnetic attraction enable us to understand, by analogy, what happens in the process of the fall.

Zerilli is obviously unfair. Analogy is a fundamental means of scientific thinking and the manner in which Angeli tries to make the analogy perfect by fitting Kepler's famous experiment into Borelli's pattern of descent is rather ingenious.[259] Let C, says he, be the center of a magnet of which the attracting pole be E, let A be an iron ball, and let them both float upon water. Then if the magnet turned around the center C and the pole E described the arc EH, the iron ball would certainly move along the arc AGH and rejoin the pole of the magnet in H. It will doubtlessly not outrun it and come to the point O, as perhaps Borelli would have it.[260]

Poor Angeli! He is, decidedly, not very good at physics and Zerilli does not miss the opportunity to point it out and to show that the analogy asserted by F. Stefano does not exist. The two cases have nothing in common: the iron ball is not drawn toward the center, but toward the pole of the magnet; *vice versa,* the heavy body tends toward the center of the Earth, and not toward the North pole.[261]

But what about the experiential and the experimental proofs alleged by Borelli? The case of the boat that conserves its speed though, after having moved in a straight line, it describes a curve? What about the cases of the circular pendulum and of the ball running around in an inverted cone?

As for the boat, Angeli thinks that it would, indeed, conserve its speed in spite of its change of direction provided its motor continues to act; but if it were separated from it, it would slow down.[262] In the cases of the pendulum and the ball exactly the same will happen: they will not conserve their speed, but will slow

---

[253] *Cf. Terze Considerationi,* 9.

[254] *Confermazione,* 9 sq.

[255] *Cf. Terze Considerationi,* 18: "Hora nel Schema del Sig. Borelli posto il graue A, nelle sommità della Torre AE, e girata questa con la Terra, il graue A descriue l'arco AFM e sempre rimira il centro per la medema linea fisica AC perche questa corre dietro al graue. Staccato il grave dalla Torre, hà in realtà impressa vna velocità, che lo porta in giro, mà vna velocità respettiva verso il Centro, qual deve sempre mentenere, e conservare rimirandolo nel medesimo modo, et per medesima perpendicolare che gli corre pur dietro," *cf.* p. 45: "Se non sopra giunger altra cagione, che lo sospinga, ò impedisse."

[256] *Cf. Terze Considerationi,* 18: *Cont.* "Per vna similitudine poniamo che il punto C, sia centro d'una sfera di calamita esquisita nella circonferenza della quale sia il polo E e dentro la sfera della sua attiuità sia collocata la sferetta, di ferro A, ma impedita con qualche ostacolo che non possa correre ad unirsi con il polo E; e questi corpi siano collocati sopra qualche piano orizontale, che posse girarsi intorne al centre C; si facino girare; il polo E, descriuera l'arco EH e il ferro A, l'arco AM maggiore e in consequenza con maggiore velocità; sia leueato ostacoló, si chè il ferro corra ad unirsi con la calamita, e nel tempo che si consuma nel farsi questa unione il polo E, sia arriuato in H."—It is interesting to note that Angeli, who has rejected the explanation of gravity by attraction (*cf. supra,* p. 368 sq.), now makes an appeal to Kepler.

[257] *Cf. Confermazione,* 17: "Ma, Padre mio, questà e la conclusione ignota replicata con altra frase."

[258] *Ibid.,* 18.

[259] *Cf. Terze Considerationi,* 19: "Per una certa similitudine posto che il punto C sia centro d'una palla di calamita il cui polo traente sia E, e dentro la sfera della sua attivita sia una palla di ferro A, ed ambedue gallegino nell'acqua, e girino intorno al Centro C, è certo che il polo B descriuera l'arco EH ed il ferro A passerà l'arco AM maggiore di quello, e pero piu veloce, e mentre il ferro A si conduce alla calamita, il polo si arrivato in H."

[260] *Ibid.,* 19: "Crede il Signor Borelli che il ferro si unirà còl polo H, o pure col punto O, per cagione dalla magior velocità circolare? Io per me credo che con polo E, il H, perche questo rimira, e all vnione con questo aspira."

[261] *Cf. Confirmazione,* 19.

[262] *Terze Considerationi,* 11: "OFRED. Ma la finitione del Sig. Borelli si reiduce à questo, come segue à dire, che quando vn mobile viene spinto da díuerse virtù motiue, ciascuna fa il suo officio, non impensendo vna l'operatione dell' altra etc."

down, the more so as their path is more curved.[263] Borelli indeed tells us that they won't. But did he, in fact, perform the experiments he presents to us as proofs of his assertion? Angeli, who obviously senses that there is something wrong in Borelli's description, is very much in doubt about it, and though the mathematician declares that this does not matter, Ofreddi is nevertheless prepared to bet that he did not make the former and finds it even more difficult to believe it concerning the latter.[264]

Angeli adds that, if Borelli's conception were right, a number of things which he deems unbelievable would follow. Thus if a shaft were made through the center of the Earth, a body thrown in would never be able to reach this center, prevented from doing it by the "corpulence" of the Earth.[265] Jupiter, in order to let Phaeton fall into the Po, would be obliged to strike him with his lightning at a time when he was still pretty far from it, and what marvelous mastery on the part of the eagle who dropped the turtle on the head of Eschyles![266]

Angeli is much better at purely mathematical or purely kinematic reasonings and his last argument against the conception of descending motion as developed by Borelli is certainly by far the strongest. The process of the fall, he objects, could conform to Borelli's theory only if it were admitted that the circular motion and the motion downward do not interfere with each other, and that therefore the heavy body (the stone falling from the top of the Tower) would move all the time with the same circular (transverse) velocity.[267]

But in this case, as the circles upon which it moves would diminish constantly and become infinitely smaller than the initial one, the descending body, before reaching the center, would have to perform an infinite number of revolutions,[268] which means that it would never arrive there.[269]

F. Stefano is right and wrong at the same time: wrong in believing that the necessity of performing an infinite number of circumvolutions would prevent the falling body from reaching the center of the Earth, and right in stating that in Borelli's conception (especially the last one, that which he developed in his *Risposta*) the falling body would never arrive at the center of the Earth, any more than planets which gravitate toward the sun would fall upon it. They would, forever, turn around.

Stefano degli Angeli's objection is presented by him in a jesting form, aimed at ridiculing Borelli—Zerilli is rightly incensed about it.[270] It is, nevertheless, serious enough to incite the latter to meet it on theoretical grounds, and even to modify accordingly—and rather unfortunately, as we shall see—Borelli's theory, or, as he puts it himself, to explain it more clearly in order to prevent misunderstandings, such, for instance, as Angeli's.

So far Angeli has dealt only with the first "circular" theory of Borelli. To tell the truth, the new one, that which replaces the circular *impetus* by the rectilinear one, the *impetus* along the tangent, appears to both him hardly worthy of a detailed treatment and criticism. In this case, the authority of the great Galileo should suffice.[271]

Father Stefano, however, is not without some misgivings. How shall Borelli react to such a detached treatment? Father Stefano is deeply concerned about it, and not without reason. As we have seen, Borelli, in his *Risposta*, has been rather unpleasant. Consequently, though it is perfectly clear that Borelli's ire has never had any foundation and though Father Stefano has already protested against the misinterpretation of

---

[263] *Ibid.*, 12, 14.

[264] *Ibid.*, 12: "OFRED. Che il Sig. Borelli habbia fatta questa esperienza mi è più duro du digerire, che della passata."

[265] *Ibid.*, 14: "OFRED. Grande infortunio sarebbe il nostro, se si mouesse la Terra, che la discesa del graue venga impedita dalla corpulenza della Terra, perche se questa girasse, e fosse forata sino al centro, e noi fossimo qui, se e vero, che come dice il Sig. Borelli.
"Anzi che forse non gl'arriuarebbe che in lunghissimo tempo, e forse mai, mentre, come dice il Sign. Borelli alla facciata 20, il moto all'ingiù si anderebbe sempre più debilitando, cio è li spatij passati non caminarebbero con la proportione delli quadrati delli tempi."

[266] *Ibid.*, 15: "OFRED. Si chè se quello, che cadesse fosse qualche animal viuo, le assicuro che hauerebbe un gran capogiro, e storidmento. CONT.: Feronte forse hauerà esperimentato quanto dice il Sig. Borelli, perche fulminato da Gioue douete principiar à piombare molto lontano dal Pò sopra il qual cadete. OFR. Gran maestria bisognara, che fosse stata quella dell'Aquila, che credendo il capo del mizero Eschilo vna picha gli lasciò cader sopra la Testitudine per romperla."

[267] *Terze Considerationi*, 36: "MATT. Concessa questa dottrina al Sig. Borelli, non credo che si possa altro dedurre, se non che girando il sasso nella sommità della Torre con la reuolution diurna, e soprauenendo il moto all'ingiù questo non impedisca quello, si che per questo capo seguiti à muouersi con la medesima velocità circolare, la quale si come prima del moto discensiuo muoue in giro, così dopo la giunta di questo sequiti a muouer in giro." As we see, and as we have already seen (*cf. supra*, p. 363), Angeli understands perfectly the meaning and structure of the theory of Borelli; yet he does not accept it

because he believes it to be a purely mathematical construction. *In rerum natura* the two motions will not be interindependent, and the transverse one will not maintain its velocity because (*ibid.*, 37): "accostandolo sempre più e più al Centro, al quale questo sempre aspira, ed anela rimirandolo constantemente, acciò questo possa fare, besogna che si muoua in giro con tanta più tardità."

[268] *Ibid.*, 15 sq.: "Supposto che il sasso girasse con il moto diurne, e cadesse sin al Centro della terra, se fosse vero che il graue trattenesse la medesima velocità, che aueua nella sommità dell'arco, perche li concentrici all'arco, che descrive il graue nella predetta sommità vicini al Centro sono infinitamento minori di quella porzione di esso descritta nel tempo della discesa, il graue prima d'arriuare al Centro girarebbe infinite volte intorno ad esso."

[269] This seems to Angeli to be so absurd a consequence that to show it implied in Borelli's theory amounts to a refutation of this latter.

[270] *Cf. Confermazione*, 24.

[271] *Cf. infra*, n. 277.

his intentions and feelings, he believes it necessary to take some precautions. Thus he writes: [272]

Though Sig. Borelli has certainly not deceived himself, and knows how falsely he has had suspicions about the sincerity of the respect that I have for him, and the esteem that I have for his great virtue and the profundity of his knowledge, it is necessary that the considerations of what he says, should be preceded by a true account of the history [of his, Stefano degli Angeli's, polemics with Riccioli, and, incidentally and accidentally, with Borelli [273]]: Already five years ago [I was studying] the measure of my infinite inverse spirals which I published only last month in my booklet, on which occasion I had the fortune to discover that the trajectory of the descending heavy body will be inside the circle, and not outside, in opposition to what the famous Galileo said and the likewise famous Riccioli in his *Almagest*. I have informed this latter by letter and I have received such an answer that I had to write to him with a greater determination. And this exchange [of views] developed in such a way that I imposed upon myself the obligation to print my theory about the measure of the infinite spirals, and in a *scholium* to touch on this

[272] *Terze Considerationi*, 22: "MATT. Hò notato anch'io cosi alla sfuggita nel leggere, questi stupori; e acciò che il Sig. Borelli se desinganni, e conosca quanto falsamente habbia sospettato della sincerità della seruitù, che gli professo, e della stima, che faccio delle sue gran virtù, e della profondità del suo sapere, è necessario, che alla consideratione de quanto dice, io facia precedere un racconto verdico."

[273] *Ibid.*, 22 sq.: "Gia anni 5 hebbi fortuna d'incontrare nella misura delle mie infinite spirali inverse, che ho publicata solo li mezi passati nel mio libretto, con la quale occasione connobi, chè la semità del graue cadente sarebbe fuori della circonferenza del circolo, e non dentro, come contro il famoso Galileo diceua l'altretanto famoso Riccioli nel suo Almagesto. Ne auisai questo con una lettera, e ne riceuei risposta tale, che mi obligò à scrivergli con maggior risolutione; e tanto s'inoltrò il negotio, che io m'impregnai seco, che hauerei stampato li miei sensi distesi la mesura della infinite spirali, e in vn scholio toccaua questa controuersa con il P. Riccioli, ma per varij impedidimenti differii la stampa del mio libretto. L'anno passato mentre mi ritrouano in Padoua per la festa del Santo capitò vn certo da Bologna, il quale mi portò à donare da parte del dottissimo Sig. Montanari vn' esemplare delli suoi Esperimenti Fisico-Mattematichi, e perche questo mi fù diretto per opera di vn suo scolare, mi mandò di più questo di propria cortesia un trattatelo manuscritto del medemo Montanari, nel quale si conteneua quella dottrina delli angoli de diuersa inclinatione da me registrata nel mio Dialogo secondo alla facciata 119 accio ne discesi il mio parere. Questa scritturetta mi riduesse à memoria l'impegno nel quale mi trouauo con il P. Riccioli; sì che venuto à Venetia principiai à scriuere li miei primi Dialoghi, li quali mentre scriuendo mi fu da parte del Sign. Borelli fatto il pretioso regalo del suo eruditissimo libro *De Vi Percussionis*, il quale non potendo per le predette occupationi studiare, come era mio desiderio, e hauendo gia scritto a Bologna al Sig. Montanari, con il quale non haueuo hauuto prima amicitia, con occasione di ringratiarlo del dono fattomi, qualche cosa circa la dottrina delle mie Spirali, questi impedito dal male d'occhi mi fece scriuere dal predetto scolaro, che il dottissimo Sig. Cassini gli haueua detto scriuer qualche cosa di questa Spirale il Sig. Borelli nel predetto libro; il che poi fù da me considerato nel mio primo Dialogo della facciata 29, e poi vn giorno accidentalmente vidi quel tanto, che mi diedi maestria di scriuere quanto hò scritto nel 2 dalla facciata 113. Questa è la pura verità, che duerà seruire à desimpressionare il Sig. Borelli, e à sodisfare alle sue merauigli, le quali il Sig. Conte può prencipiar a rappresentare."

controversy with F. Riccioli, but because of various obstacles I have postponed the printing of my booklet. Last year, when I found myself once more in Padua for the feast of the Saint, I encountered a certain person from Bologna, who brought me, as a gift, from the most learned Sign. Montanari [274] a copy of his Physico-Mathematical Experiments, and as this [book] has been addressed to me through the medium of one of his pupils, he sent me moreover, by his own courtesy, a small manuscript treatise of the same Montanari, in which was contained the theory concerning the angles of diverse inclination which I reported about in my Second Dialogue on page 119 where I said, too, what I thought about it.[275] This small work recalled to my memory the obligation I had toward P. Riccioli; so that when I came back to Venice I started writing my first Dialogue, [and] while I was writing it I received on the part of Sig. Borelli the precious gift of his most learned book *De Vi Percussionis*, which, on account of the above-mentioned occupation, I could not study as it was my intention; and as I had already written to Bologna to Sign. Montanari, with whom before that I did not stand in close relation, something about the theory of my spirals, in order to thank him for the gift he made to me, this latter [Montanari] prevented by eyesore [from writing himself] has made his above-mentioned pupil write [to me] that the most learned Signor Castelli has told him that Sig. Borelli has written something about these spirals in his above-mentioned book, which was hereafter examined by me in my first Dialogue on page 29; then, on a certain day I by accident noticed[the implications of his theory] which gave me the incentive to write what I have written on the page 113 of the second. This is the pure truth. . . .

Thus concludes Father Stefano, in the pious hope that truth will prevail and will dispel the painful impression which Borelli might have had on seeing himself attacked by him.

We can now proceed to the examination of Borelli's new theory. As usual, it is the Count who is entrusted with the exposition and it is the mathematician who presents the criticism: [276]

CONT. In order to explain his new conception, he [Borelli] reminds us of a certain *natural property of the circular motion*, which has the faculty of removing the moving bodies from the circumference upon which, at first, they moved, moving them along a straight line, tangent to the circle at the point of separation, every time it occurs that the moving body detaches itself from the said circumference, etc. He says then that this is commonly received as certain and evident.

[274] Geminiano Montanari, born June 1, 1633, in Modena; died October 13, 1687, in Padua. Doctor of Philosophy, Law, and Medicine; first lawyer in Florence, then Astronomer of the Grand Duke of Toscana and Mathematician of the Grand Duke Alfons IV of Modena; later Professor of Science in the University of Bologna and finally Professor of Mathematics in the University of Padua.

[275] This theory of Geminiano Montanari having nothing to do with the problem of fall has not been reported by me.

[276] *Terze Considerationi*, 31: "Cont. Per esplicare questo suo nuouo pensamento, ricorda *quella natural proprietà del moto circolare*, che hà facultà d'estrudere, e allontanare i mobili dalla circonferenza, nella quale primo si moueuano per una linea retta tangente il cerchio nel punto della separatione, qualunque volta accada, che il mobile si spicchi dalla detta circonferenza etc. Dice poi che questa cosa è communemente riceuuta come certa, e euidente." *Cf. supra*, p. 375.

But, answers Angeli: [277]

MATT. This theory is no by means commonly accepted and, in this generality, is false. Heavy bodies [which] gyrate or are gyrated on the circumference of a circle either tend towards its centre by their own gravity, or they do not. When they gyrate in the first manner, as it would occur in the case of the Earth, this rotation does not have the virtue of extruding. Read Sig. Galileo in Dialogue 2 of the *Systema Cosmicum*, p. 137, of the Latin edition, and see with what evidence he demonstrates this truth. The wheels that extrude are those toward the centre of which the extruded body does not tend by its own gravity, and such are all our wheels, the Earth excepted, because in this [one] there is a struggle between motions toward diverse directions, but not in the others, as Kepler has very well shown it in the first book of the Epitome of the Copernican Astronomy, p. 137, which combats the arguments of the *impetus* and the extrusion.

According to Angeli, Borelli knows it quite well. Yet he commits the error of believing that, after the separation of the heavy body from the wheel which turned it around, this circular *impetus* of the said body, will, in all cases, degenerate into a rectilinear one, which is by no means the case. The tendency of the body toward the center of the Earth, which prevents it from being "extruded" by the rotating Earth, will also prevent the circular *impetus* from degenerating into a rectilinear one.

It is rather curious that, whereas Father Stefano did not misunderstand the "circular" theory of Borelli, he, obviously, has been unable to grasp the meaning of the rectilinear one. He seems to believe [278] that Borelli's conception implies an *actual* motion along the tangent, that is [because of the very great speed of the transverse motion of the Tower compared with the very great tardiness, at the beginning of the fall, of the motion downward] an *actual* removal of the stone from the surface of the Earth and its projection upward.

Zerilli, thus,[279] is quite right in explaining that this is by no means the case, and that—though a sufficiently great velocity of the Earth's rotation could produce such an "extrusion" just as it could neutralize completely the force of gravity—the factual velocity is too small. The motion by the tangent is, therefore, only a virtual one; yet, it has to be treated as real, just as the

downward motion due to the gravitational *impetus* of the body.

Father Stefano protests, too, against Borelli's device of supposing the moving body deprived of gravity. Is it not a common experience that one can throw a stone much farther than a piece of cork? If the body in question were completely deprived of gravity, it would not be able to receive any *impetus,* and thus, would not be able to move at all.[280]

Zerilli once more is right in replying (a) that it is by no means true that the action of a given *impetus* is facilitated by the weight of the body, a very heavy body being very difficult, or even impossible, to move; (b) that to deprive, practically, a body of its gravity is very easy: a body moving on the surface of the Earth, always at the same distance from its center, or a body equilibrated on a balance, is freed from the action of gravity; (c) that the operation proposed by Borelli is a purely intellectual one.[281]

The discussion between Borelli and Angeli can be considered as perfectly sterile; indeed, it does not bring us any nearer to the solution of the problem of the trajectory of the body falling from a high tower on a rotating Earth. But it is, at the same time, extremely revealing. It shows the extreme difficulty of performing the precise degree of abstraction, halfway between mathematics and empiricism, which constitutes the peculiar pattern of thinking of modern theoretical physics.

Stefano degli Angeli cannot think as a physicist; or at least, only seldom. He cannot—without becoming a pure mathematician—forget that he is dealing with natural bodies, that these are heavy and bulky, that they are attracted by the Earth, or tend toward its center, that they fall through the air which resists their motion. Bodies—if they are not purely mathematical bodies, in which case they obey purely mathematical laws and, descending towards the center, describe necessarily an inverted spiral—do not move in a vacuum, but *in hoc vero aere,* and therefore (a last argument, which, in Father Stefano's opinion, gives a final refutation of Borelli's conceptions)[282] cannot maintain the speed of their transverse motion,

[277] *Ibid.,* 31: "MATT. Questa dottrina non è altrimente communemente riceuuta, e in questa generalità è falsa. Li graui che giri ò sono girati per la circonferenza di vn circolo al centro del quale sijno portati dalla grauita propria ò nò. Quando girano nel primo modo, come accaderebbe nel moto della Terra, questa rotatione non hà virtù di estrudere. Lega il Sig. Borelli il Galileo nel Dialogo 2 del Sistema Cosmico nella pag. lat. 137 e vererlà con quanto euidenza dimostri questa verità. Le ruote che estrudono sono quelle al centro del quali l'estruso non è portato dalla propria grauità, e queste sono tutte le nostre ruote, eccettuata la Terra, poiche in quella vi è la pugna di moti verso diuerse parti, non in questa come benissimo dice il Keplero nel libr. 1 dell'Epitome dell'Astron. Coper. alla facciata 137, il qual combattimento caggiona quell'empito e estrusione."
[278] *Ibid.,* 34.
[279] Cf. *Confermazione,* 32 sq.

[280] Cf. *Terze Considerationi,* 35 sq.; cf. *supra,* n. 232. Stefano degli Angeli cannot distinguish the *weight* of a body from its *mass.*
[281] *Confermazione,* 36.
[282] *Terze Considerationi,* 36: "MATT. Ma perche la discesa auicina il graue sempre più, e piu al centro, e lo constringe vrtare in parti del mezo mosse con minor, e minor velocità, e in consequenza più resistenti, a rintuzanti il moto circolare del graue, quindi ne segue che per accidente il moto del graue all'ingiù impedisca il moto circolare. Lo impedisca anco per accidente perche accostandolo sempre più, e più al centro, al quale questo sempre aspira, e anela rimirandolo constantemente, acciò questo possa fare bisogna si muoua in giro con tanta più tardità. Questo è il mio pensiero, nè meno io mi posso dar ad intendere altrimente. Non concedendo adunque al Sig. Borelli questo suo fondamento, non posso concedergli li altri suoi discorsi sopra esso fondati. Conuengo con lui nella conclusione, cio è che il

because the descent brings the heavy body always nearer and nearer to the centre, and constrains it to strike into the parts of the medium moving with a lesser and lesser speed, and consequently resisting and opposing more and more the circular motion of the heavy body, it follows therefrom that *per accidens*, the motion of the heavy body downward hinders the circular motion. It hinders it *per accidens* because, as it approaches more and more the centre to which it tends all the time and in the same way, it is necessary, in order that this should be done, that it should move in a circle with a correspondingly greater tardity. This is what I think, and I cannot possibly conceive it in another way. Not conceding therefore to Sig. Borelli this his fundamental [assumption] I cannot concede to him his other assertions based upon it. I agree with him concerning the conclusion that the falling heavy body will always be on the perpendicular to the horizon, though I believe that it will be on it in a more precise manner than according to him; yet I believe it not because of his new reasons, but because of the old, and rancid ones, asserted first by others, and by me received from them.

Thus Father Stefano degli Angeli.[288] Let us now turn our attention to Diego Zerilli's restatement and "elaboration" of Borelli's doctrine.

To begin with, Zerilli exposes, once more, Borelli's conception. In doing it, he disregards, for the time being, as Angeli had done, the new shape given to it by its author in the *Risposta*, that is, the replacement of the circular *impetus* by the rectilinear one.[284] On the other hand, he substitutes for the straightforward reasoning of Borelli a *quasi*-infinitesimal treatment of the movement of the descending stone, insisting upon the necessity of considering the progress of its motion, or motions, not during large intervals of time, such as one second of the hour, but only during very small ones, such as a sixth or even a tenth of a second.[285]

He does it, obviously in order: (*a*) to "save" the continuity of the "oblique" motion, and (*b*) to prepare the reader for the curious (and erroneous) infinitesimal conclusions of his own theory. Angeli's analysis of the downward motion is, once more, rejected.[286]

It could be objected that the composition of these two motions [287] can also be conceived somewhat differently, supposing the point *A* [fig. 24], in so far as it is the *terminus a quo* of the descent *AE*, to be transported on the arc *BA* [to the point *B*], wherefrom the ball would be conduced to the point *D* by the descending motion; it would

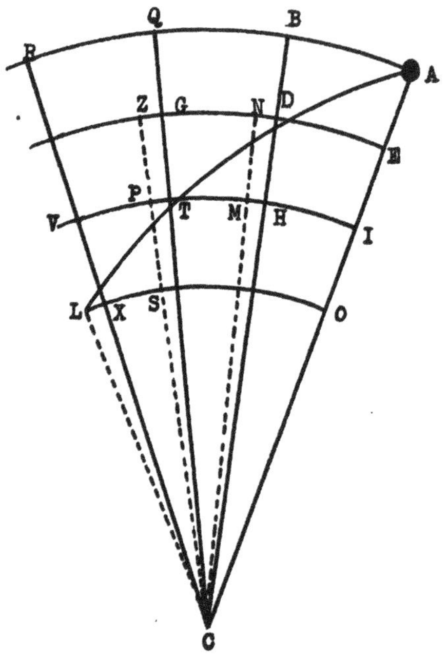

FIG. 24

follow that the ball would not abandon the semidiameter *AC* transferred to *BC*. But in order to show the weakness of this discourse, let us suppose it to be true, that is [let us suppose] that at the end of the first time, the ball really finds itself joined to the point *D* of the semidiameter *AC* transported to *BC*, and let us consider what will follow in the subsequent times. The ball, finding itself in *D*, must likewise perform there its two motions, that is to transport the point *D*, *terminus a quo* of the descent, on the circumference *DG*; and because it is supposed that the transverse *impetus* is *quasi* the same as at the beginning,[288] it would have to traverse an arc *DZ quasi* equal to *BQ*,[288] and consequently, greater than *DG*: hence the ball falling along the semidiameter *ZC* will find itself beyond the semidiameter *AC*, transported in *QC*, by the whole interval *TP*. And the same will occur later on. Thus, in every case, in this composition of motions the ball will always have to outdistance the semidiameter upon which it begins its motion. But that the oblique motion cannot be reasonably compounded by the transverse on *AB* and the descending on *BD*,[289] seems to [result] convincingly from the fact that the ball *A* does not, in truth, move on the circumference *AB*, from which it departs, but runs transversely on other, inferior, circles concentric to *AB*, one of which is *EN*. If, therefore, on all these circles the ball has to conserve, in all the minimal times, *quasi* the same degree of transverse *impetus*, which cannot produce less than what is required by its energy, it is necessary that the ball, detached and free, and not conjoined with the semidiameter *AC*, should traverse the

---

graue cadente sempre sia nella perpendicolare all'Orizonte, anzi io credo che sia in essa più presisamente di quelle lo pone lui, mà non già per le sue nuoue ragioni, ma per le vecchie, e rancide assegnate prima da gli altri, e da me da essi riceunte."

[283] *Ibid.*, 42: "Ofred. Io quantunque stij attentissimo più che non soglio fare à qual si sia intricata demonstration geometrica, o resolution analitica nulladimeno confesso ingenuamente la mia ignoranza, che non intende queste consequenze, cio è come si deduchino dalle premesse."

[284] In doing it he follows the example of Borelli himself who, as we have seen (*cf. supra*, p. 371) begins his *Risposta* by restating the theory sketched in the *De Vi Percussionis*, and only then adds the correction.

[285] *Cf. Confermazione*, 25.

[286] *Ibid.*, 25 sq.

[287] The descending and the circular motions.

[288] *Quasi* . . . Zerilli has admitted (*cf. supra*, p. 378) that the resistance of the air alleged by Angeli will actually slow down the motion, yet only in an exceedingly small degree.

[289] As Angeli wanted it to be.

space EN, *quasi* equal to AB, and hence larger than ED. Thus it will be true that the *terminus a quo* of the transverse motion will be conduced downward on the semidiameter AC. For the same reason, the terminus, or starting point of the oblique motion, will descend on the semidiameter NC till M, and the starting point of the oblique motion PL will descend on the semidiameter PC till S, and thus always.

Curious reasoning. Zerilli (who seems to have been seduced, and misled, by Borelli's device of determining the position of the falling body at the end of each "time" by representing it as moving, or as having moved, towards the center of the Earth from corresponding imaginary starting points) objects to Angeli's determination of the trajectory of the descending body as compounded by a motion *first* on AB, etc., and *then* on DC (BD), and substitutes for it his own: *first*, the starting point moves down on DC (BD), and *then* on AB (EN). The two manners of compounding are by no means equivalent and Zerilli, quite rightly, goes on: [290]

From this it follows that the said *terminus*, or starting point of the transverse motion, while it is descending more and more toward the center, comes nearer and nearer to the site of the semidiameter from which it starts the said oblique motion, just as the point N, descending toward M, will be continually approaching the semidiameter AC, the more the ball moves downward toward M, and this occurs because of the convergence of the semidiameters the mutual distance of which diminishes the more they approach the center. In the same way, the said *terminus*, or starting point of the transverse motion in S, will be nearer to the semidiameter AC than in P, and so on. So that approaching to the center C the said *terminus* will be separated from the semidiameter by a distance of only less than a finger breadth, or than a hair breadth, though the highest circumference ABR, which corresponds to it, be equal to the quadrant of the terrestrial equinoxial circle.

Thus, in virtue of this continuous retrogradation of the *terminus a quo* of its transverse motion, the falling body, in spite of its continuous outrunning of the corresponding radius would, nevertheless, be drawn nearer and nearer to it. In this way the danger revealed by Angeli—that of the body turning around the center without ever reaching it—would be luckily avoided.

Alas! Zerilli does not see that his ingenious modification of the Borellian theory is nothing more, or less, than Angeli's conception turned backwards, and he pursues proudly: [291]

From all this result some astonishing things, *first* that the true transverse motion of the ball A, accomplished on circles parallel to AB, will be always equally rapid, traversing equal spaces in equal times, and that all [these spaces] together will be equal to the circumference of the great circle ABR, that is, equal to the space which the ball would traverse if it remained on the top of the semidiameter AC; this because of the minimum space EN being equal to AB, of the space MP equal to QB, and of the space SL equal to QR, the same being true of all the other subsequent

[290] *Cf. Ibid.*, 28.
[291] *Ibid.*, 28 sq.

smallest little spaces. It follows therefrom that all the transverse motion that the ball A will have made on the oblique way ANPL will be equal to the motion AR, which it would have made staying on the top A of the semidiameter. Still, the removal of the ball from the semidiameter AC, with which it was connected at the beginning of its motion, will be much smaller than the motion of the summit A of the semidiameter AC on ABR, because MI is smaller than NE or indeed than BA,[292] and PM is equal to QB [293] and therefore the distance PI from the semidiameter AC is smaller than QA, and likewise the distance LO will have to be smaller than AR. Moreover, though for a good stretch [of time] these distances will be increasing—thus PI is smaller than LO—yet they will diminish toward the center, and finally will be reduced to nothing; [294] the reason for this extravagance is not that the *impetus* and the transverse motion of the said ball on the minimal arcs EN, MP, SL, etc., are continually weakening, or slowing down, but that the distances of the *termini a quo* or of the starting points of the transverse motions [from the semidiameter] shrink more and more, and that they approach more and more the semidiameter from which the motion starts in virtue of the convergence of the said semidiameters.

Having thus succeeded in reconciling Borelli's theory of the constancy of the *impetus* with the necessity of bringing the falling body back to the radius-vector from which it departed, Zerilli concludes: [295]

It does not seem to me that one should make much fuss about the fact that the said intervals PI, LO, etc., from the first semidiameter AC, will *per accident* be diminished and shortened, because the ball will not be moved on the said transverse spaces IP and LO, but on the minimal transverse arches EN, MP, SL, etc., which, as it has been said, always move equally rapidly with the summit A of the semidiameter AC, and therefore the *impetus*, and the true transverse motion of the ball will not be at all retarded. We can therefore concede freely that the distances of the ball from the semidiameter AC which are PI, LO, etc., are always gradually diminishing in the measure that they approach the center; it is enough for us that the *impetus* and the transverse motion by which truly the ball is moving be either not at all, or only very slightly, weakened, and that therefore the said ball would outdistance somewhat the semidiameter upon which it starts to move and with which it should have to gyrate together. From this it follows, in virtue of that small deviation, that the curve ANPL will finally be drawn to the center, but it will still not be a spiral, as P. Angeli has wanted it. From this it follows likewise that a stone and other things falling, in the hypothesis of the gyration around the center, will not have to turn around as many times as P. Angeli believed. And I point out that when Sig. Borelli, in his letter, has said that the falling ball should, in this hypothesis, outrun

[292] The MI, as well as the NE is composed of a great many minimum arcs of which the *termini a quo* are somewhat displaced to the right.
[293] The *impetus* being the same, the arcs, or their lengths will be equal, cf. however, pp. 382 and n. 288.
[294] Thus the moving body will be brought back to the semidiameter and to the center of the Earth.
[295] *Ibid.*, 29 sq. The falling body will thus describe a curve bulging out at the middle and then drawn back to the center of the Earth. Nevertheless, as Galileo wanted it, its trajectory will be equal to the path it would have described if it did not fall at all, but remained on the top of the tower, indeed an astonishing thing!

the foot of the very high tower transported to BC by a space smaller than a finger breadth, he has said it in order to make a kind gesture and give every advantage to the opponent, but that in truth it will be much smaller, provided the last examined theory were admitted.

XIII. S. DEGLI ANGELI CONTRA ZERILLI

The "strategic" situation in which Stephano degli Angeli found himself after Zerilli's *Confermatione* was closely parallel to that which resulted from the publication of Michele Manfredi's *Argomento*. Once more it was a disciple who counter-attacked in the name of his master; and once more Father Stefano who, obviously, enjoyed the opportunity, was able to deal with the new and unknown opponent in a manner that he could not use toward such a well-known and respected scholar as Borelli, making it clear, at the same time, that it was the master, rather than the pupil, who was the true recipient of the blows.

"Zerilli! Diego Zerilli." [296] Stefano degli Angeli does not hide the fact that he doubts the very existence of such a person, thus [297] Ofreddi:

The name is truly majestic and breathes I don't know what air of gravity; but the cognomen is most capricious, and perhaps even mysterious. I don't know whether there is in the world, or whether there ever has been anybody who was, or is, named thus. But this is of no avail, because Signor D. Diego confesses that the theories, which he sets forth, have been received by him from the mouth of the most learned Sign. Borelli, his master; replying to him we shall give satisfaction to both; imitating in this way those who exorcise the possessed; because as the Devil speaks through their mouths, the exorciser, by beating the possessed, hurts the other and torments the Devil.

As for the Devil, that is Borelli himself, Father Stefano still bears him a grudge. He cannot forget that Borelli accused him of plagiarism. He wants to make it quite clear that it was Borelli and not himself who started the polemics in his letter to Michel Angelo Ricci and that, moreover, when reading the famous passage of the *De Vi Percussionis* in which Borelli criticized the spiral theory of fall,[298] he had good reasons to believe himself being attacked. "Indeed," says the Mathematician,[299]

[296] *Cf.* Stefano degli Angeli, *Quarte Considerationi*, 10. Ofreddi makes a pun: Zerilli is a diminutive of Zero. "*Ofr.*: è il suo cognome in certo mode diminutivo del Zero." In the *Quarte Considerationi*, S. degli Angeli—or his printer—spells *Matematica* (with one *t*) and *Ofredi* (with one *d*).

[297] *Ibid.*, 2: "OFRED. Il nome e veramente maestoso, e spiro un non sò che di gravità, il cognome par mi molto capritioso, e forse anco misterioso. Non sò però se hora viua alcuno al mundo, ò sia mai vi sciuto, che così fosse, ò sia nominato. Ma questo poco importa, poiche confessando il Signor D. Diego che le dottrine, che adduce le ha rittrate della bocca del Dottissimo Signor Borelli suo maestro, rispondendo ad esse, sodisfaremo ad ambidue: imitande in cio quelli li quali scongiurano l'inspiritati, poichè parlando il Demonio per bocca di questi, battendo il scongiutante l'inspiritato, percuote questo, e tormenta il Demonio."

[298] *Cf. supra*, p. 358. *Quarte Considerationi*, 3.

[299] *Quarte Considerationi*: "MAT. Se deuo dir il vero, quando ho letto queste parole del Sig. Borelli, ho dubitato che parlasce

If I must speak the truth, when I read this passage of Sign. Borelli, I thought [at first] that he spoke about me; because, though I had not yet printed anything concerning this subject, it was already some years since I had informed some of my friends (and perhaps Sign. Borelli himself [who was] one of the dearest) of having the geometrical measure of certain infinite spiral spaces, one of which was the space that would be described by the path of the heavy body falling, if the Earth moved in diurnal motion only, and [if it fell] in the plane of the Equator. Yet, in the letter sent to the most illustrious and most learned Sig. Michel Angelo Ricci he declared who those moderns were against whom his criticisms have been directed, and said it was Sign. Fermat, who already some time ago had considered this line, as P. Mersenne reports it in his *Phenomenis Balisticis*.

As for the contents of Borelli's remarks in *De Vi Percussionis*—as well as later—Father Stefano cannot but repeat what he already developed in his *Considerationi*. It is not the "moderni," criticized by Borelli, who "do not abide by their presuppositions"; it is Borelli, who misunderstands them. Indeed, they admit that the heavy body will move along the semidiameter. Borelli, on the contrary, believes that its circular *impetus* will be conserved, and that it will outrun the semidiameter, a theory contrary both to sound reason and to observation.[300]

Zerilli objects, it is true, that Borelli's conception cannot be opposed for purely experiential reasons. It is a rigorously mathematical theory, that must be treated as such. *Ad sensum*, the trajectory described by the falling body in the first four seconds of its motion—and our experiments do not extend farther—would be indistinguishable from the vertical, because the distance between these two lines would be too small to be observable.[301] Nonsense, replies Father Stefano. *Ad sensum*, as a matter of fact, there is nothing to be observed. If the Earth moved, we would move with it, and the transverse motion being common to the tower, the falling body and ourselves would be as if it were not. It is true, on the other hand, that purely mathematical objects are not, and cannot, be realized in the physical world. Yet, it is precisely *that* which Borelli forgets when he tells us something that Zerilli seems to have overlooked, i.e., that the circular *impetus* will be conserved in the case of *real* fall.[302] Moreover Zerilli con-

meco; poiche se bene non haueuo ancora stampato cosa alcuna in simil proposito, erano pero alcuni anni, che haueuo date parte ad alcuni miei amici (e forse al Sig. Borelli medemo vno delli più cari) di hauer la misura geometrica di certi infiniti spatij spirali, vno delli quali sarebbe il contenuto dalla semita del graue cadente, quando la terra si mouesse con il solo moto diurno, e ciò nel plano dell'Equatore. Egli però nella lettera diretta all'illustrissimo, e Dottissimo Sig. Michel d'Angelo Ricci dichiara quali fossero questi moderni, contra li quali erano indirizzate questi parole, cio è il Sig. di Fermat, che hauena già un pezzo fà considerato questa linea, come riferisce il P. Mersenno nelli suoi Fenomeni Balistici. Il che per hora si osserui." *Cf. supra*, p. 371.

[300] *Ibid.*, 4.

[301] *Ibid.*, 6 sq.

[302] *Ibid.*, 8, 13.

tradicts himself. In his *Terze Considerationi* Angeli had asserted that the transverse motion of the falling body will be retarded by the resistance of the air. Yet Zerilli objected that this was not sufficient, and that what was required was not a retardation in general, but a retarding cause acting *with mathematical precision.*[303]

But let us turn away, for a while, from Zerilli and the discussion about things that would happen, or would be observed, or would not be observed, if the Earth moved. It is obvious, by the way, that all the disputants, Stefano degli Angeli as well as Borelli and Zerilli, are convinced that the Earth does move. Let us come back to the fundamental opposition between Borelli and Angeli—the former asserting that the circular *impetus* conserves itself during fall, the latter, on the contrary, that it will be "extinguished"[304]—and to the experiments alleged by Borelli in order to prove his thesis.

We have seen[305] that, reporting on these experiments, Father Stefano expressed some doubts about their having been actually performed. Now he has proof that they were not and even could not be made. When expressing his doubts, he had, indeed, asserted that the question was not important, and that even *if* the experiences were *true*, they would not prove anything. Nevertheless, it is obvious that he is highly pleased in being able to demonstrate that Borelli's experiments were not only "untrue," but not even real. What a pity, says Ofreddi, that he did not make a bet! He would have won.[306]

Father Stefano confesses that in the demonstration of Borelli's substitution of imaginary experiments for actual ones,[307] he was helped by his newly acquired friend, Montanari.[308]

MAT. The most learned Sig. Geminiano Montanari, mathematician at the University of Bologna, having read these our controversies wrote me a very modest letter, in which he told me that he thought that the experience of

Sign. Borelli was an obvious paralogism,[309] but that he had so great an esteem for his genius, that he did not dare to assert it, and asked me for my opinion. The error that he believed [he found] in Sig. Borelli being this: *BA*, *BG* [fig. 20] being two pendulums, the lengths of these must be in reciprocal proportion to the squares of their vibrations, that is, of their numbers. Thus, if *AB* is four times *BG*, then in the time in which *BA* describes one circumference, *BG* must describe two, and not four as Sign. Borelli seems to have experimented.

When I came to Padua last October [continues Father Stefano][310] the same objection had been made to me by the most learned Sig. Carlo Rinaldini,[311] then first philosopher in this University. Indeed, when we came to the experience, we have manifestly seen [that] the pendulums *BA*, *BG*, moved horizontally, have the same periods, and are subjected to the same laws of the pendulums as if they were moved in the vertical plane; from which we already recognized the error of Sign. Borelli; and that Sig. Ofreddi was right in guessing that he [Borelli] had produced an experience not made by him.

Having thus given honor to whom honor was due, Stefano degli Angeli proceeds to the analysis of the Borellian experiments:[312]

MAT. Let us, gentleman, note two things that we have observed. The first that when we conferred an *impetus* on the ball *A*, this [ball] did not describe a circumference of a

[303] *Ibid.*, 16; cf. *supra* p. 378.
[304] *Ibid.*, 12, 15.
[305] *Cf. supra*, p. 379.
[306] *Ibid.*, 15.
[307] As we know, it is a rather frequent case in the XVIIth century. And even later.
[308] *Ibid.*, 15: "MAT. Il Dottissimo Sig. Geminiano Montanari Matematico nello studio de Bologna hauendo letto queste nostre controversie mi scrisse una lettera molto modesta, nella quali mi diceua dubitare che l'esperienza dal Sig. Borelli fosse vn'euidente paralogismo, ma che stimaua tanto il suo valore, che non osaua affirmarlo, ma che mi pregaua del mio parere. Il sbaglio che credeua nel Sig. Borelli era, che essendo *BA*, *BG*, doi pendoli, le lunghezze di questi sono in reciproca proportione delli quadrati delle loro vibrationi, cio e del numero. Hora essendo *AB*, quadrupla della *BG*, nel tempo che *BA*, descriue una circonferenza, *BG* ne deue descriuer due, e non quatro come pare hauer esperimentato il Sig. Borelli." Signor Montanari is slightly mistaken: the laws governing the motions of the perpendicular and of the circular pendulums are not identical. This however, has no bearing on his objection to Borelli's "experiments."

[309] The expression used by Montanari is rather characteristic: he treats the experiment as a logical exercise.
[310] *Ibid.*, 15: "Venuto à Padoua l'Ottobre passato mi fù fatta la medema obiettione dal Dottissimo Sig. Carlo Rinaldini hora filosofo Primario in questo studio. E venendosi all' esperienza, vedesimo manifestamente li pendoli *BA*, *BG* mossi Orizontalemente hauer li medemi periodi, e soggicere alle medeme leggi di pendoli come se fossero mossi in piani verticali: onde conobi manifestamento lo sbaglio del Sig. Borelli, e che il Sig. Ofredi haueua indouinato, che hauena addotto vn esperienza da esso non fatta." Stefano degli Angeli, as we see, makes the same mistake as Geminiano Montanari.
[311] Carlo Rinaldini, scion of an old Siennese family, was born in Ancona in 1616. Having studied mathematics, philosophy, and theology, he was appointed by Urban VIII "superintendent of the militia" of the pontifical armies; Innocent X entrusted him with building fortifications in the region of Ferrara. In 1649 he became first lecturer of philosophy in the University of Pisa, from 1663 to 1667 taught philosophy to Cosme III (Medici), and in 1667 went to Padua as professor of philosophy at the University. Member and several times even president of the Accademia del Cimento, Carlo Rinaldini enjoyed a high reputation among his contemporaries (Montanari, Borelli, Viviani, Magalotti, Bullialdus, etc). On Rinaldini, cf. Charles Patin, *Lyceum Patavinum*, 52 sq., Palavii, 1682.
[312] *Ibid.*, 17: "MAT. E notino loro Signori due cose che habbiamo osservato. La prima che secondo conferimo l'empito alla palla *A*, questa non descriueua circonferenza di circolo, ma più tosto ellisse, tutto pero grandi, o piccole in tempi eguali. La seconda è, che quando tiranimo la palla nel sito *G*, descriuendo il suo circolo, o ellisse non si conseruaua nel medesimo Cono *BAE*, ma ampliaua più la sua circonferenza. Cosa che sarebbe stata molto gradita dae Sig. Borelli quando fosse stata da esso esperimentata, perche si vede manifestamente che trattiene buona parte dell'empito, che haueua in *A*; perche douendo per ragion del pendolo nel medesimo tempo che descriueua l'*ADEA*, descriuere due nel sito *G*, essendo la velocità, o empito maggiore di quello, che è necessario à descriuer le due circonferenze *GF*, s'allarga dalla linea *BA*, e fà più ampij li giri."

circle, but rather ellipses, all of them, large and small, in the same time.[313] The second, that when we pulled the ball to the position G, [this ball], describing its circle, or ellipse, did not maintain itself on the cone *BAE*, but increased its circumference. Something that would be taken in account by Sig. Borelli if this experiment had been made by him, because it clearly shows that if [the ball] retains a good part of the *impetus* that it had in A, according to the proportion of the pendulum, it must, in the same time in which it described [the circle] *ADEA* describe in the position G two [circles], and as its speed or its impetus is greater than is needed for describing the two circumferences *GF*, it moves farther from the line *BA* and makes larger circumgyrations.

Despite his errors concerning the circular pendulum, Father Stefano is perfectly right. The centrifugal force —for the same linear velocity—is inversely proportional to the radius of the circular motion. Thus, for the small circle *GF* it will be four times as large as for the large one; wherefore the ball of the pendulum will not remain on the conical surface previously described by its thread (*BA*), but will increase its distance from the perpendicular.

How far will it remove itself? Stefano degli Angeli does not know exactly.[314] In any case, it will not reach the distance that it should reach in pure theory. The real conditions of the experiment are to be reckoned with. Thus he concludes: [315]

MAT. That it move so far away from *BA* that the two circumferences become equal to *ADEA* and to four times *GF*, I do not believe. Indeed the pull downward, the resistance of the air, and other things must progressively diminish the velocity that it had in A on *E*.[316]

The case of the ball running on different levels of an inverted cone is, obviously, exactly similar to that of the pendulum. The experiment described by Borelli is impossible. The ball will not stay on the lower level describing a smaller circle. If it is not slowed down but, as Borelli maintains, retains its *impetus* (linear velocity) it will return to its former position. Yet

strangely enough Father Stefano does not criticize this second, Borellian "experiment," though, already in the *Terze Considerationi*, he has declared it even more doubtful than the pendular one. He does not mention it at all.

I must confess that I am unable to explain this silence. It is hard to believe that Father Stefano could miss the analogy between the two cases. It is just as hard to believe that he would disregard the opportunity to "rub in" a blunder of his so highly revered friend.

Let us now turn back to Zerilli. Father Stefano reproaches him for misrepresenting his, Angeli's, views, and particularly for not understanding that the problem they are discussing is not an abstract, kinematical one, but deals with real bodies falling on a real Earth. Thus he says,[317]

MAT. I don't know whether I should consider Sig. Diego to be blind or to be malicious, because he quotes my words printed on page 18 and not those which are on page 17,

where Father Stefano tried to explain the physical mechanics of the process. Indeed, continues the mathematician,

demonstrating the proposition that the falling heavy body will be maintained on the same perpendicular physical line, I assigned this reason for its perseverence: "namely, that the mass of the Earth is spherical, is physically obvious;

---

[313] The discovery that the circular pendulum will describe ellipses as well as circles had been made by Robert Hooke already in 1666. It is not impossible that Father Stefano had been informed about Hooke's experiments in the Royal Society. In any case the error of ascribing to the circular pendulum the property of being isochronous is entirely Stefano degli Angeli's and results probably from his—erroneous—belief in the isochronism of the perpendicular pendulum and the—erroneous, too—identification of the laws of the circular and the perpendicular pendulums.

[314] This is a difficult question and implies the knowledge of the law of centrifugal forces, a knowledge that at this time no one possessed, but Huygens and Newton, neither of whom had published his findings. Even R. Hooke did not know the correct answer; *cf.* my: An unpublished letter of Robert Hooke to Isaac Newton, *Isis*, 1952.

[315] *Ibid.*, 17. "MAT. Se poi s'allarghi tanto che le doi circonferenze sijno eguali alla *ADEA*, e quadruple dalla *GF*, io non lo credo, mentre il tiramento all'insu, l'impedimento dell'aria, e altre cose deuono più, e più debilitare la velocità che haueua in A ouero *E*."

[316] Stefano degli Angeli is right: the radius of the circular pendulum will not be twice *GF*.

[317] *Cf. Quarte Considerationi*, 19 sq.: "MAT. Io non sò se debba giudicare il Sig. Diego per cieco, ò malitoso, mentre recita le mie parole registrate a carte 18 e non quelle contenute alla faccia 17. Dimostrandone il senso che il graue cadente viene trottenuto nel medesima linea physica perpendicolare, di questo trattenimento, o conservatione n'assegno questa cagione. 'Che la mole terraquea sia sferica fisicamente è manifesto; e si assegna communemente per più valida cagione, che tutti li corpi, che la compongono desiderano vna intima vninone fra se, la qual conseguiscono nel miglior modo che possono, componendo vn corpo di figura, per cosi dire, più vna di tutte le altre, quale è la sferica. Dà cio ne segue, che tutti anelino, et aspirino al centro delle cose graui, con il quale hanno un tal rispetto. Quindi pure ne segue che collocate in vn sito di questo gran corpo sferico non si lascino cacciar da quello che con violenza, ma se se gl'apre l'addito di potersegli più accostare, subito precipitino verso quello. Da questo parimente ne segue, che per ipotesi girata la terra circa il proprio centro, le parti ad ella contigue, e anelanti all'unione con il centro con quella virtù, che noi chiamiamo grauità, girino parimente con essa, rimirando constantemente da quella parte che sono il centro per la via più breve, che è quella linea, che conjunge il graue e il centro.' Et esemplificando nello schema del Sig. Borelli soggiungo le parole sopradette alla facciata 18 (staccato il graue etc.). Hora Sig. Zerilli mio, che il graue cadente habbia vna velocità, che lo mantenga nel medesimo semidiametro, ma perche tutti li corpi, che compongono la mole terraquea desiderano vna intima frà se. Questa è l'originaria causa di tutti li effeti, che sopra ho recitati, buona poi, o cattiua che sia. Il graue cadente è conseruato nel medemo semidiametro fisico, perche le parti componenti la terra desiderano vn' intima vnione per la via più breve mediante quella virtù, che noi non sapendo cosa sia, la chiamiamo grauità. La qual causa anco // quando non fosse buona, non per questo il mio discorso *petit principium* et prova una cosa per se stessa, ma solo addurrebbe une causa insufficiente, o falsa."

and as the most valid proof of this [sphericity] is [the fact] commonly given that all the bodies which compose it desire an intimate union (among themselves), which they pursue in the best manner they can, composing a body of a form [that is], so to say, more *one* than all the others. From which it follows that all [bodies] tend and aspire to the center of the heavy things with which they have this relation. It follows therefrom that, located in a certain place of this large spherical body, they will not let themselves be separated from it, unless by violence, and that, if, later, they have the possibility to come nearer to it, they will immediately precipitate themselves toward it. It follows therefrom likewise that if, by hypothesis, the earth were rotating around its own center, the parts contiguous to it, and tending towards this center with that virtue which we call gravity, will equally rotate with it [the Earth], tending [in the same time] constantly toward the center [and striving to reach it], from whatever place they are, by the shortest way, which is that line which joins the heavy body and the center." Then, taking as an example the scheme of Sign. Borelli, I added the abovementioned words on page 18: if the heavy body (grave) were separated etc.[318]

Thus, my dear Sig. Zerilli, there is a reason why the falling heavy body will have a velocity that will maintain it on the same semidiameter; it will occur because all the bodies which compose the mass of the Earth desire an intimate union. This is the original cause of all the effects that I have described *supra*, be they good or bad. The falling heavy body is maintained on the same physical semidiameter, because the parts which compose the Earth desire [to reach] an intimate union by the shortest way, [driven] by that virtue, which, not knowing what it is, we call gravity. Thus, even if this cause were not good, my discourse would not therefore be a *petitio principii* and prove a thing by itself, but only be [guilty] of assigning an insufficient or false cause.[319]

Zerilli's—and of course Borelli's—conceptions are based, as we have seen,[320] on the Galilean idea of the body's indifference toward motion and rest. According to them a body, once put in motion, retains its *impetus* and does not *spend* it in the production of a horizontal translation. The body, moreover, subjected to a circular motion (on a rotating Earth) acquires through this motion a rectilineal impetus, and it acquires it, in any case, even if, *de facto*, it is not projected ("extruded") by the centrifugal force, and does not separate itself from the Earth. Let it be separated from its support, as in the case of the free fall, and the horizontal *impetus* will reveal itself by exercising an influence upon the trajectory described by the body.

Stefano degli Angeli, of course, denies it:[321]

MAT. [If] The heavy body *A* [is] let free, [and] if the excess of the impetus overcomes the tendency of the heavy

body towards the union with the point *E*, there is no doubt that the heavy body will separate itself from the perpendicular *AE*; but as it does not overcome it, just as the excess of speed of a superior point does not overcome the speed of an immediately inferior point, it follows therefrom that the heavy body descends on the perpendicular *AE*.

Borelli's conception is once more presented: [322]

CONT. Sign. Zerilli will say that these theories are *gratuitously* asserted; indeed, he adds that the heavy body will "willingly stay in whatever place of the said [spherical] surface it be transported, because they are equally removed from the center of the Earth. And because of this reason the swimming piece of wood, and a crystal ball rolled upon a horizontal surface are seen to be indifferent, and not to have any resistance against the transverse motion, and able to be transported by any minimal force, because it is of no importance for them to be in one place rather than in another, because it is no more distant from the center of the Earth than the first one."

It would seem that there is nothing to object to Zerilli's reasoning. And yet: [323]

MAT. I do not sell these [my] theories for more than they are worth. Yet they seem to me reasonable, because they enable us to explain a phenomenon, which is really observed, namely, that the heavy body moves along a physical line perpendicular to the horizon, whatever be the altitude from which the motion starts, be the Earth in motion, or, as it is really, at rest. I believe it to be certain that the heavy body will willingly stay in any place whatever on the surface of the Earth to which it would be transported, because it would be equally distant from the center. Yet I believe that when it is in a place, it will have a certain repugnance to be transported into another; and that, therefore, a force proportionate to its bulk will be needed for transporting it. And I do not see, though Sign. Zerilli sees it, namely that a floating piece of wood and a crystal ball in the horizontal plane will be indifferent and

---

[318] *Cf. supra*, p. 378.
[319] *Cf. supra*, p. 378.
[320] *Cf. supra*, p. 373.
[321] *Quarte Considerationi*, 21: "MAT.: . . . Lasciato il graue *A*, in sua libertà se l'eccesso dell'empito *A*, superasse l'appetenza dell'unione dell graue con il punto *E* non vi ha dubio che stacherebbe il graue dalla perpendicolare *AE*; ma non la superando, si come non supera l'eccesso di velocità d'alcun punto superiore la velocità del suo punto inferiore immediatemente, guindi ne segue che il graue discenda per la medesima perpendicolare *AE*."

[322] *Ibid.*, 31 sq. "CONT. Dirà il Sig. Zerilli queste dottrine sono *gratis asserite,* mentre egli soggiunge che il graue 'stara volontieri in qualcunque luogo della (della) superficie sia trasportato purche sia egualmente remoto dal centro terrestre. E per questa cagione il legno gallegiante, ed una palla di cristallo intorna la superficie Orizontale spianata si vede esser indifferente, e non hauer resistenza nessuna al moto trasversale potendosi da ogni minima forza trasportare non importando gli punto star più in vn luogo, che in vn altro perche non si scosti più di primo del centro della terra.'"

[323] *Ibid.*, 22: "MAT. Io non vendo queste dottrine per più di quello che vagliono. Mi paiono pero ragionevoli, mentre procurano render ragione di un fenomeno, che realmente appare, cio è che il graue si muoue per la linea fisica perpendicolar all' Orizonte da qual si voglia altera principij il moto, si muoua ò sia immobile come è realmente la terra. Io credo di certo che il graue starà volontieri in qual si sia luogo della superficie terrestre, nel quale sia trasportato, purche sia egualmente distante dal centro. Ma quando e in vn luogo, ad esser trasportato nell'altro credo che vi habbia qualque repuganza; che percio a trasportarlo vi voglio vna forza proportionata alla sua corpulenza. Ne io uedo quello, che vede il Sig. Zerilli, cio è che il legno gallegiante, ed vna palla di cristallo intorne l'Orizontal spianata sia indifferente, e non habbia repugnanza alcuna, si che ogni minima forza sia sufficiente a muouer la trasuersalmente, purche stia sempre egualmente lontanata dal centro; ma stimo più tosto, che quando il graue è in quiete, à rimouerlo da quel luogo vi voglia vna forza considerabile."

will not have any repugnance [against being moved], so that a minimal force would be sufficient to move them transversely, because they would remain always equally distant from the center. I rather think that when a heavy body is at rest, to remove it from a certain place you need a considerable force.

Poor Father Stefano! He is obviously right: in order to move a large, heavy body you really need a considerable force. We must not blame him, at least not too strongly, for not understanding how to reconcile empirical data with the Galilean principle of inertia, and for advocating, at least tentatively, the return to Keplerian conceptions: [324]

CONT. You touch upon a theory that I have had several times in my imagination. I have seen serious philosophers, whom I revere in the innermost of my heart, say that the heavy body is most indifferent to this circular motion equally removed from the center. In fact I have always had some doubts concerning this theory, because it seemed to me that, being a corporeal thing, a body must have a certain inertia with respect to local motion, and that it cannot be moved if not by some force, even if the hindrance of the resistance of the medium were removed; and that the mere fact of corporeality is the reason for this difficulty of being moved. And thus it seems to me that though the whole terrestrial globe, as such, were not heavy, and gravity were only that of its parts; nevertheless, to move it from place to place, would be needed a considerable, and [even] a large force, I say large compared to those forces which, with us, transport heavy bodies from place to place. Thus it seems to me that in order to make the great bodies of the planets move, certain considerable force would be required, even if the ether were much more yielding than it is, because, after all, they [the planets] are bodies.[325] The reason why I think so is that I have not much liking for the theory of Aristotle according to which the motion in the void is instantaneous, but rather believe that it cannot be performed except in time. And if this has some

point of reason, I will say: If a heavy body falling naturally to the center through the void, that is, conducted by the force of its gravity, cannot traverse a space unless in time, why shouldn't we say that a body cannot be transported to a place, to which it has no inclination but only an indifference, unless by a proportionate force? [As for] the minimal force [326] I don't understand [what] it [is] but it rather seems to me that a force is required such that a lesser one would not be sufficient.

The conception developed by Angeli explains very well why the falling body remains on the radius connecting it with the center of the Earth. The *impetus* alleged by Zerilli—and Borelli—cannot remove the body from this radius: it is too weak. Yet, even if it were not, Zerilli's objections would be worthless, for, according to him, the *impetus* in question is not circular, but rectilinear, and, if the Earth rotated, it would move the falling body not on a circumference, around the center, but on a tangent, that is away from it.[327]

Zerilli's objections against the conclusions drawn by Stefano degli Angeli from Borelli's theory, namely that the falling body will have to perform an infinite number of gyrations before arriving at the center and thus will be incapable of ever reaching it [328] are just as bad as everything else in his pamphlet. As for his attempt to make fun of Father Stefano: [329]

CONT. Sign. Zerilli considers as fables these his [Angeli's] deductions and this infinite gyration around the center; and before explaining the alleged vanity of these assertions on p. 24, he says: "It is just as contradictory as it is absurd that a circumference of the circle, because of its being adjacent to the center, should be infinitely smaller than the arc of the most external circle of the wheel. Indeed, it is known that two determined quantities of the same kind are in a finite proportion. I will though, against the strict meaning of the words, interpret them in a good sense."

MAT. I am much obliged for the courtesy of Sig. Zerilli who forgives me this error. Yet I believe I was

---

[324] *Ibid.*, 22 sq.: "CONT. V. S. tocca vna dottrina, che più volte mi è passata per la fantasia. Ho veduto grauissimi filosofi, li quali reuerisco con il più intime del mio cuore, dire che il graue sia indifferentissimo à questo moto circolare egualmente remoto dal centro. Io realmente ho sempre hauuto qualche dubio sopra questa dottrina, mentre parmi, che l'esser vna cosa corporeale sia vna certa inertia al moto locale, si che non possa esser mossa che con qualche forza, leuati anco li impedimenti della resistenza del mezo; e cosi che il solo esser corporeo sia cagione di questa difficolta ad esser mosso. E cosi mi pare che se bene tutto il globo terrestre come tale non fosse graue, e che la grauità sia delle sole parti; nulla dimeno à muouerlo da luogo à luogo sia necessaria forza considerabile, e grande; grande dico paragonata à quelle forze, che appresso noi trasportano li graui da luogo à luogo. Cosi parmi che à far muouer li gran corpi delli pianeti si richieda qualche forze considerabile, anco quando l'etere fosse molto più cedente di quello che è, perche alla fine questi sono corpi. La ragione poi del mio dubitare è, perche à me mai é piaciura la dottrina d'Aristotile, che il moto nel vacuo fosse instantaneo, ma più tosto stimo che non potesse esser fatto che in tempo. E se questo hà punto del ragioneuole dico cosi. Se il graue cadente naturalmente al centro per il vacuo, cioè condotto dalla forza della sua grauità, non puo passare vn spatio che in tempo, perche vorremo dire che vn corpo possa esser trasportato al un luogo al quale non ha inclinazione ma solo indifferenza se non da forza proportionata?

[325] According to Zerilli—and Borelli—a body does not oppose any resistance to motion.

[326] According to Zerilli, Borelli and Galileo, the smallest force will move the largest body.

[327] *Ibid.*, p. 25.

[328] *Ibid.*, p. 27.

[329] *Quarte Considerationi*, 29 sq.: "CONT. Il Sig. Zerilli reputa per favole queste sue dottrine e questo giramento infinito intorno al centro; e prima d'esplicare la vanità pretesa delli suoi asserti alla facciata 24 dice queste parole: 'Ne meno se gl'oppone quanto sia assurdo che vna circonferenza di cerchio per esser vicina al centro debber esse infinitamente minore d'un arco del cerchio estremo della ruota. Essendo noto che due quantità terminate del medemo genere hanno proportion finita. E voglio contre l'espresso significato delle parole interpretarlo in buon senso.'

"MAT. Io resto molto obligato alla cortesia del Sig. Zerilli, che mi condonna questo errore. Io però mi credeuo di parlare con vn geometra Capomastro, quale è il Sig. Borelli, che sò che si sarebbe vergognato à notare la frase ordinaria di parlare che assume l'infinita per l'indefinito. Mà cosa hà Sig. Ofredi che ride cosi saporosomente. OFR. Rido, perche mi arrecordo di vn modo molto galante adoptato dal Galileo per leuarsi simil nota. Lotario Sarsi nella sua Libra Astronomica riprese il Galileo di simil modo di parlare; al quale altro non propone nel Saggiatore, se non che esso saluasse il detto del sauio, che *stultorum infinitus est numerus*, essendo certo che tutti li huomini che sono, e saranno, sono contenuti in numero finito, e limitato."

speaking to an accomplished geometer, such as Sign. Borelli, who knows that it would be pedantic to reject the common way of speaking which uses the infinite for the indefinite. But for what has Sign. Ofredi to laugh so contentedly?

OFR. I am laughing because I am reminded of the elegant manner adopted by Galileo in order to dispose of a similar remark. Lotario Sarsi, in his *Astronomical Balance* reproached Galileo for a similar way of speaking; to which the former in the *Saggiatore* did not reply anything but that he abided by the *dictum* of the Sage that *stutorum infinitus est numerus*, though being certain that all men that are, or will be, are in a finite and limited number.

This terminological error, or levity, being corrected—at Zerilli's expense—Angeli lets the Count reproduce his reasoning as presented in the *Terze Considerationi*,[330] and lets the mathematician reformulate it, stating that the body will never arrive at the center because it will have to make a *quasi* infinite number of revolutions.[331]

Repetition of old arguments, restatement of his old views—as a matter of fact, it is nearly all that Stefano degli Angeli is able to do. The "falseness" of the Borellian experiments, the ineptness of Zerilli's attempt to come to the rescue of his master, have reconfirmed him in his old, fundamental, convictions. His opponents deal with the rotation of the Earth—or at least, try to do so—as if it were a simple "case" of rotation, not different, in principle, from any other "case." That is precisely their error. The case of the Earth is *different* from all the others, and centrifugal forces that are encountered in all the other cases, are not produced by this one. It could be different, of course. God coùld have made the Earth rotate in such a way that it would "extrude" bodies and send them flying away. But,[332]

MAT. I, for the present, do not want to investigate what God, and nature, could have done. I know well that, *de facto*, there is no extrusion, because I see that a heavy body, even a small one, remains at rest on the top of the tower, and, in descending is always on the same physical perpendicular. I believe also that if the Earth rotated, even much quicker than in 24 hours, it would not extrude [it] by that [rotation], because I do not admit this circular motion to have an extrusive faculty. Therefore, if it be true that necessarily it must extrude, then, every time that Sign. Zerilli will prove it, he will prove at the same time that the Earth, most certainly, does not move.

---

330 *Cf. supra*, p. 379.

331 *Quarte Considerationi*, 30 sq.

332 *Ibid.*, 37: "MAT. Io per hora non voglio ricercare quello che posse fare Iddio, e la natura. Io so bene, che *de facto* non vi è estrusione, perche vedo che il graue ben picciolo stà immobile sopra la cima della torre, e discendendo e sempre nella medesima perpendicolar fisica. Credo anco che quando girasse la terra, anco assai più veloce delle hore 24 non per questo estruderebbe, mentre non capisco che questo moto circolare habbia facoltà estrusiua. Se poi è vero che de necessità debba estrudere, ogni volta che il Sig. Zerelli prouerà insieme che la terra certissimo non si muoua. Ma veniame à quella conclusione, che hò detto di sopra cauarsi dalla dottrina del Sig. Zerelli, dalla quale aparirà questa verità."

But a theory which would enable us to demonstrate *that the Earth does not move* is obviously, for Father Stefano, disqualified for that very reason: it cannot be true. Indeed, it is quite possible that Stefano degli Angeli's opposition to Zerilli's as to Borelli's conceptions stems, to a considerable degree, from his Copernicanism. He sees them, curiously enough, as allies of Riccioli.[333]

Thus, the assertion, in Zerilli's and Borelli's dynamics, that the rotation of the Earth produces centrifugal forces cannot be true: it is contrary to fact. Moreover, as Angeli has already mentioned elsewhere, this theory implies, as a necessary consequence, another, equally false and non-existent, phenomenon,[334] that is that the descent of the heavy body will be the slowest on the Equator, and infinitely more rapid on the pole.

Indeed, in spite of the fact that the Earth, as a whole, turns around its axis in twenty-four hours and that therefore bodies, wherever they be placed, accomplish this circumgyration in the same time, they do not' move with the same speed. In modern terms, we have to distinguish between the angular and the linear velocity of different points of the Earth's surface, and consequently, recognize that a body placed on the Equator will move with greater tangential velocity than those that are placed elsewhere.[335]

MAT.: Thus the heavy body does not move in a circular motion with the same degree of speed on all the parallels, though describing them all in 24 hours, but the circular motion on the Equator will be a good deal quicker than that on the other parallels, and on the Pole there will be none. Wherefore in the plane of the Equator the great degree of circular speed will strongly hinder the descensive one, on the other [parallels] less and less, and on the pole it will not hinder it in any way; therefore the heavy body will fall less [quickly] on the Equator than on the other [parallels] and on the pole [it will fall] with the greatest speed.

Stefano degli Angeli's reasoning, with its mixture of correct deduction and the refusal to accept the result, is rather characteristic of the backwardness of the Italian science of his time. Elsewhere, in France or in England, a man of the value and the abilities of Father

---

333 Indeed all of them teach, against Galileo, that if the Earth moved, bodies would move differently from the way they would if the Earth were at rest.

334 *Ibid.*, 37: "Quest'è che il graue discenderebbe all'ingiù tardissimo nel piano dell'Equatore, e infinitamente più veloce sotto il polo nello sfera parallela." Father Stefano is, once ʼmore, betrayed by his lack of precision. The falling body will fall on the poles with a greater velocity than on the Equator, or anywhere else, but this velocity, though the greatest, will not be infinite.

335 *Ibid.*, 37, 38: "MAT. Hora il graue in tutti li paralleli non si muoue di moto circolare con il medemo grado di velocità, ma descriuendoli tutti in 24 hore, il moto circolare nel piano dell' Equatore è assai più veloce, che nelli altri paralleli, e nel Polo e nulle. Adunque nel piano dell'Equatore il gran grado di velocità circolare impedirebbe molto il descensiuo, nelli altri meno, e meno, fine che nel Polo non impedirebbe in conto alcune. Adunque il graue descenderebbo meno nei Equatore che nelli altri, e sotto il Polo con velocità grandissima."

Stefano would not deprive the Earth of the "faculty of extruding"; and would have announced to the world—before Hooke and Halley—that heavy bodies fall more slowly on the Equator than on the Pole.

As a matter of fact, even the distinction between linear and angular velocity and the role ascribed to the former—though already used by Borelli in his *Theorica Planetarum Mediccarum* for the explanation of the elliptical orbits of the planets [336] seems to be too radical for F. Stefano. Thus it is attacked by Ofredi as contradicting the very basis of Borelli's own theory.[337]

OFR. I do not know what Sign. Zerilli would reply: I know, though, what I would say; [namely] that these degrees of velocity are not sufficiently diverse for hindering diversely [the descent]; this because of a certain uniformity that they possess, namely that all the parallels are described in 24 hours. In the scheme of Sign. Borelli, though the two circles GC and AE are different, and though the point A revolves with a greater velocity than the point G, nevertheless, these two unequally rapid circular motions, are not different in [their power of] extrusion because they are performed in the same time. Indeed, the same circumgyration GC performed in a shorter time and consequently with a greater degree of velocity would have a greater energy for extruding and hinder the descensive motion. But in the parallels, to whatever extent the degree of velocity of the one be greater than the degree of velocity of the other, nevertheless, there is no diversity in the extruding [action], because they all gyrate in the same 24 hours. They would of course extrude and hinder more [strongly] if they were made to gyrate in a shorter time. And in order to answer more briefly, and to give a stricter reply, I distinguish two greater degrees of velocity; [that] the greater degree of circular velocity, [greater] because the circle [is described] in a shorter time has the virtue of extruding in a greater degree, 1 concede; [that] the greater degree of velocity [greater] because of the amplitude of the larger circle, but gyrated in the same time, has a greater faculty to extrude

[336] Cf. my paper quoted *supra*, n. 146.

[337] Cf. *Quarte Considerationi*, 38: "OFR. Io non sò cosa sia per respondere il Sig. Zerilli; sò bene che direi che questi diversi gradi di velocità non sono di quelli sufficienti ad impedire diversamente, per certa loro uniformita, cio è che tutti li paralleli sono descritti in 24 hore. Nel Schema del Sig. Borelli quantunque li due circoli GC, AE siano diuersi, e che il punto A, giri con maggior velocità del punto C, nulladimeno questi due moti circolari inegualmente veloci non sono differenti nell' estruder, perche sono fati nel medemo tempo. Bene la medema circulation GC, fatta in minor tempo, e in consequenza con maggior grado di velocità hauerebbe maggior energia d'estrudere, e impedire il moto discensiuo. Cosi nelli paralleli all' quantuncque il grado di velocità d'vno sia maggiore del grado di velocità dell'altro, nulladimeno non vi è diuersita nell'estrudere, perche tutti girano nelle medeme hore 24. Bene più estruderebere, e impedirebbero quando si facessero girare in tempo minori. E per respondere più brevemente, e dar vna riposta stringata, distinguo due maggior grado di velocità; il maggior grado di velocità circolare perche il circolo in minor tempe hà virtu d'estrudere maggiormente, concedo; il maggior grado di velocità per ampiezza di maggior circolo, ma girato nel medemo tempo hà magior facoltà d'estrudere, e impedire, lo nego. Ne occore che V. S. noti la diuersita del schema del Sig. Borelli, oue li circoli sì ampiano per l'allungamento del lato BC, in BA, e nel caso delli paralleli per l'ampliatione dell'angolo al centro della terra, e vertice del Cono, perche è quasi il medemo."

and to hinder, I deny (negate). It happens that you have made the distinction between the scheme of Sign. Borelli where the circles increase because of their elongation from the side BC, in BA, and the case of the parallels [which become greater with] the increase of the angle toward the center of the Earth or of the vortex of the cone, because these are *quasi* the same [case].

Stefano degli Angeli's distinction of the two kinds of "greater velocity" is subtle—too subtle, alas, for it means, in modern terms, that the *linear* velocity of the circumgyrating body has no importance for, and no effect on, the centrifugal force arising from its motion: it is the angular velocity, and only this one, that counts.

Poor Father Stefano! His ideas about the centrifugal force are rather hazy (but Borelli's were not much better, if they were better at all).[338] Thus he is able to conclude, Borelli's teaching notwithstanding: [339]

MAT. The Equator and all the parallels, because [they] are all described in the space of 24 hours have the same faculty to extrude and to hinder the motion downwards.

XIV. RICCIOLI CONTRA STEFANO DEGLI ANGELI

Stefano degli Angeli tells us that he had just concluded the composition of his sixth dialogue and was ready to send it to the printer when he became aware of the publication, by Riccioli, of an *Apologia* directed against him. He had, therefore, to postpone the printing and add to the sixth dialogue a seventh. Thus he announces to the reader: [340]

I had already composed the sixth Dialogue, when, on the very day that I wanted to start the printing, I became aware of the *Apologia* of the most learned P. Gio. Battista Riccioli, printed in Venice, in favor of his Physico-mathematical argument against the Copernican System, in a good part of which he attacked my *Prime* and *Seconde Considerationi*. And though the errors, and falsehoods that it contains could easily be recognized by you, if you had seen my said *Considerationi*, nevertheless I have judged it worth while to announce them in the seventh Dialogue in order that you may know them with less effort.

Father Stefano is rather glad to add that the *Apologia* of Giambatista Riccioli is directed not against him alone, and that in his opposition to the Ricciolian argu-

[338] By comparison, we can appreciate the immense value of Huygens'—and Newton's—treatment of the problem.

[339] *Ibid.*, 38: "MAT. L'Equatore con tutti li paralleli, perche tutti descritti nel spazio di 24 hore, hanno la medema faculta d'estrudere, e impedire il moto descensivo."

[340] Stefano degli Angeli, *Quarte Considerationi*, Al Lettore, Padoua, 15 Oct. 1669: " . . . Havevo già disteso il sesto Dialogo, quando il medemo giorno, che voleuo principiare à farlo stampare, hebbi notitia dell' Apologia del Dottissimo P. Gio. Battista Riccioli, stampata in Venezia à fauore di quel suo argomento Fisicomatematico contro il Sistema Copernicano, in buono parte della quale pretende impugnare le mie Prime, e Seconde Considerationi. E se bene le Bugie, e Menzogne, che contiene possono esser facilmente conosciute da Te, se hauerai veduto le mie dette Considerationi, nullàdimeno ho giudicato bene auertirle nel Settime Dialogo, acciò le posse conoscere con minor fatica."

ments he has—Riccioli says so himself—some very precious allies: [341]

The Apology of the most learned S. Gio. Battista Riccioli is not directed against us alone, but as he states on pages 4 and 5, against several great scholars who did not like this argument; these are the most learned Juius Turrini, Jean Domenic Cassini, Geminiano Montanari, Giovanni Alfonso Borelli, and to these he has also joined ourselves.

As a matter of fact, neither of these two works—Riccioli's and S. degli Angeli's—enriches the discussion by new and important ideas. One could even say that, in some sense, they do not add anything to the debate. They close the curve of the polemics and bring us back to its starting point, they repeat and restate their arguments, but they do not improve them. The interest of these works is purely historical.

Riccioli, of course, has not been convinced—or even shaken—by Stefano degli Angeli's criticism. Just as before, he is unable to understand the meaning of the Galilean relativity of motion; just as before, he is convinced of the validity of his argument, and intensely proud of having discovered it. Yet, honest as he is, he admits that his conviction is not shared by everybody, not even by all his friends and colleagues. He even informs us—and Angeli—how many of these men of high standing and reputation as well as of unimpreached orthodoxy failed to see the light. In his *Apologia* he is at his best, learned, thorough—*doctissimus et eruditissimus*—and at his worst, irremediably committed to his system of obsolete physical conceptions, incapable of the effort of changing, or even modifying them.[342]

The point at issue being not so much the *validity* of the Copernican system, a question which for all true Catholics is decided once and forever by the infallible decision of the church, as the possibility of giving a rational refutation of this system, Riccioli starts by giving a very excellent short prehistory of it.[343] He

continues by stating the importance of the debate, which involves questions pertaining to astronomy, physics, and theology, and which became therefore a *causa celeberrima,* and tells us about his own contribution to the discussion, namely: [344]

... in the book IX of the *Almagestum Novum,* sect. iv, where from chap. 5 to chap. 34 we have presented forty-nine arguments in favor of the diurnal and of the annual motions of the Earth, and the immobility of the Sun in the center of the Universe, adjoining to them, however, their solutions. As for the immobility of the Earth, we have presented seventy-seven arguments.

Among other arguments against the system of a moving Earth, that one appeared to me as physico-mathematically evident, which I have deduced from [the fact that] the real force of percussion becomes so much greater and greater as the body descends naturally from a greater altitude, which force of percussion, *caeteris paribus,* implies a real and notable increment of the *impetus* acquired through the motion, as well as a real acceleration of the descending heavy bodies; But in the hypothesis of the motion of the Earth, this acceleration would be only [an] apparent [one], and, taken absolutely, [it would be] non-existent; [taken] physically, it is demonstrated to be so small, that it is of no importance in relation to the force of the percussion, whether the descent of the graves is imagined to occur along the circular line designated by Galileo, or along another kind of curvilinear path, for in-

velocissimam sedem? circa nos Deus omnia, an nos agat? Quandoquidem diurnam horarum 24 revolutionem, Telluri circa sui centrum, ac axem vertigine continua Oriente versus circumeunt: adscripserant olim *Heraclides Ponticus, Ecphantus Pytagoricus* et *Nicetas Syracusanus,* vt referunt Cicero lib. 2 Academ. sq. et lib. 1. Tusculanae ac Plutarchius lib. 3 de Placitis Philos. cap. 13. Praeter diurnum autem, annuum quoque motum, quo centrum Telluris circo Solem in Mundi centro quiescente circumferatur diebus 365. et fere diei quadrante, inuexerunt Aristarchus Samius et Philolaus vterque Pytagoricus attestante id Archimede in Arenario, Plutarcho 3 de Placitis cap. 11, Ptolemaeo lib. 1. Magnae constitutionis cap. 5 et 7 et indicante Aristotele 2 de Caelo textu 72. Horum opinionamenta, postquam Nicolaus Copernicus Toronensis et Canonicus Varmiensis veluti ex diuturno somnio excitauit, nouisque hypothesibus substruxit."

---

[341] *Ibid.,* 46: "Non e stato ferito contro noi soli, ma come dice alle facciate 4. e 5. contro diversi gran Letterati, à quali non è piaciuto quell'argomento. Sono questi li Dottisiimi Guilio Turrini; Gio. Domenico Cassini; Geminiano Montanari; Gio. Alfonso Borelli; Adriano Hasut; e à questi hà accopiato anco noi altri." For Stefano degli Angeli, accused by Michele Manfredi of crypto-copernicanism, this announcement must have been rather helpful.

[342] We must not be too severe: to change one's mind is difficult, and Riccioli, in 1669 is an old man. Besides, the change involved is a very radical one.

[343] G. B. Riccioli, *Apologia,* 2: "Cap. I. *De controversiae huius celebritate, ac gravitate apud Astronomos, Physicos, et Theologos, deque Occasione iterum ac Saepius de illa disserendi nobis oblata.*
"Merito inter nobiles de Natura quaestiones Seneca libro 7 cap. 2, hanc in priuis scitu dignissimam censuit, cum dixit : *Illo quoque pertinebit hoc excussisse, vt sciamus vtrum Mundus, Terra stante circumeat; an Mundo stante, Terra vertatur? Fuerint enim qui dicerent. Nos esse, quos rerum natura nescientes ferat, nec coeli motu fieri Ortus, et Occasus, sed ipsos nos oriri, et occidere: Digna res est contemplatione vt sciamus, in quo rerum statu simus: pigerrimum sortiti, an*

[344] G. B. Riccioli, *Apologia,* 2: "In libro illo 9 Almagesti Noui, Sect. 4 ubi a cap. 5 ad 34 produximus quadraginta nouem Argumenta pro Telluris diurno et Annuo motu, Solisque quiete in centro Vniversi, subiunctis tamen eorum solutionibus. Pro immobilitate autem Terrae protulimus septuaginta septem Argumenta." *Cf. Apologia,* 3: "Inter alia itaque argumenta contra Systema Terrae motae, illud mihi Physicomathematicè visum est euidens, quod deduxi e reali percussionis validitate maiori et maiori, quo ex maioris altitudine grauia naturaliter descendunt, quae validitas caeteris paribus reale ac notabile incrementum impetus per motum acquisiti supponit, adeoque realem accelerationem grauium descendentium: At in hypothesi motus Telluris accelerationem illam mere apparentem forè, et absolutè falsam; aut phyisice tam exigaum demonstratur, vt nullius sit considerationis, respectu validitatis, qua fit percussio, siue descensus grauium fingatur fieri per lineam circularem à Galileo designatam, siue per aliam speciem curuilineae tramitis, puto parabolicam, spiralem aut helicoidem; Proinde illum argumentum prius contra Galilaeum retorquendum duxi lib. 9 Almag. Noui sect. 4, cap. 19 assumpta ex ipso via circulari, esto illam talem non esse in progressu, nec vniversaliter ostendissem cap. 17 postea vero in Astronomia Reformatae lib. I Appendice 2 illud argumentum instauraui supposita quacunque via descensuo curuilineae Copernicanae hypothesi congruente."

stance, a parabolic, a spiral, or a helicoid. Accordingly I have first, in book IX of the *Almagestum Novum*, sec. 4, chap. 19, used this argument against Galileo assuming, as he does, that the body will move along a circular path, though, in chap. 17, I have shown that it is not the case for the progressing body, nor is it the general case; and later on, in book I, App. 2 of the *Astronomia Reformata*, I presented this argument supposing the trajectory of the descent to be any curvilinear line compatible with the Copernican hypothesis.[345]

Alas, in spite of being "physically evident," this argument met with some resistance. Riccioli, thus continues: [346]

It is true that there were some very learned persons, endowed with a very acute mind, who have endeavored to refute this argument of mine, either in private discussions, or in letters and dissertations that they wrote to me, or in published opuscules; others, moreover, proposed to me various objections to be solved. Among these there are to be named in the first place the Most Excellent D.D. Franciscus Hyacintus de Simiana, the Marquess del Pianizza

---

[345] Cf. *supra*, p. 354.

[346] *Ibid.*, 3: "Verum non defuere Viri doctissimi, et acutissimo ingenio praediti, qui vim argumenti huius mei cum eneruare conati sunt qua verbis coram, qua scriptis ad me litteris vel dissertationibus, aut opusculis in lucem editis. Aliqui autem dubia nonnulla mihi soluenda proposuere. Hos inter propcipuis fuere Excellentiss. Franciscus Hyacinthus de Timiana Planitiarum Marchio, de cuius insigni pietate, multiplici et profunda doctrina, in Dedicatoria meae Geographiae Reformatae perpauca, sed prae innita ipsius modestia multa dixi: qui tamen postea mei argumenti vim agnouit probauitque. Eodem verò tempore Excell. D. Doctor Philosophiae, Medicinae, Matheseos, juris et graecarum litterarum consultissimus D. Julius Turrinus Niceae in Prouincia natus plenissimam subtilitatis dissertationem ad me scripsit, et sic inscripsit.

Nicetas Orthodoxus

Seu de Controverso Mundi Systemate

Sensus Christiano Philosopho dignus

Cuius diatribe, post quinque Conclusiones, vnica tandem Conclusione hac terminatur: *Solem Diurno et Annuo motibus circumuolui, Terram quiescere ferme teneo, infailliliter credo, et aperte profiteor, non Physicis, non Mathematicis ad id impetus Rationibus, sed mero imperio fidei, Auctoritate scripturae, et nutu Romanae Sedis, cuius effata distante Spiritu veritatis, vt universos docet, mihi iurata sunto. Et post meam Responsionem Apologeticam, rursus insurrexit Duplica ad me scripta, adhaesitque pristinae suae optinioni. Posteà verò de hoc argumento mihi non semel disceptatio fuit cum amicissimis et Excellentiss. Matheseos Professoribus in Bononensi gymnasio, D. Io. Dominico Cassino, et. D. Geminiano Montanario, Operibus in lucem editis iam in Italia, et extra notissimis, sed mihi e familiaribus colloquijs longe notissimis et aestimatissimis. Subinde autem Excellentiss. D. Jo. Alphonsus Borellus in Patria Messanensi, et deinde in Pisana Academia Mathematum Professor, primum scripto, et postmodum evulgato Libro De Vi Percussionis, Propositione LVIII meum argumentum contra Telluris motum extenuare conatus est. Nouissime denique R. Fr. Stephanus de Angelis in Patauino Gymnasio Mathematicarum Professor, duplici dialogo vtramque Argumenti mei formam inefficacem esse contendit; cumque illi meo consensu respondisset D. Michael Manfredus; iterum Fr. Stephanus contra hauc responsionem insurrexit, et D. Adrianus Nasut Gallus in suis observationibus pag. 49. Hinc factum est ut veritatis discernendae studio, sedulo et accuratius hoc negotium agressus sim, et ex multis ea selegerim, quae controversiae huic decidendae mihi visa sunt opportuna."

about whose great piety, varied and deep learning, I have, in the dedication of my *Geographia Reformata*,[347] said very little, and yet much more than his modesty would bear; who, however, later on, recognized and approved the strength of my argument. And at the same time, the very excellent D. Doctor of Philosophy, of Medicine, of Science, of Law and of Greek letters, the most wise D. Julius Turrinus,[348] born in the province of Nicea, wrote to me a dissertation full of subtlety, to which he gave the title:

Nicetas Orthodoxe

or on the Meaning of the disputed System of the World worthy of the Christian Philosopher.

---

[347] *Geographia et/Hydrographiae/Reformatae/libri duodecim . . . Ad illustriss.et excellentiss.D./D.Carolum Emmanuelem/ A Simiana/Marchionem Liburni etc. . . . Bononiae, 1661.*

Charles Emmanuel de Simiane, Marquess of Pianezza, Livorno, Montecapretto, Castelnuovo, Roati and Mareto, was the son of the well-known Carles Emmanuel Philibert Hyacinte de Simiane (1608-1677), colonel-general of the armies of Victor-Amedeo I of Piedmont, first minister of the duchess of Savoy Mary-Christine of France, minister of Charles Emmanuel II. The *Geographia Reformata* was republished, in 1672, by Jean La Noue, who, however, changed the dedication: *Geographiae/et/Hydrographiae/Reformatae/. Nuper recognitae, et Auctae/Libri Duodecim :Ad Illustriss.et Excellentiss. Liberum Baronem et Dominum/D.Bartholomaeum Bertoldum/ Austriaca in Aula/Serenissimi Ferdinandi Caroli Archiducis Austriae, Olim primum Status Consiliarium, atque Supremum/ Serenissimi/Archiducis Sigismundi Francisci Cancellarium. Venetiis . . . , 1672.*

[348] Julio Torrini was born at Lantisque in the county of Nice (date unknown). Very famous as doctor and mathematician in the first half of the seventeenth century he occupied in Turin the positions of first physician of King Emmanuel II, royal librarian and professor at the University of Turin where he taught mathematics and medicine. In acknowledgment of his merits the city of Turin granted him the rights of citizenship. Later he became professor at the University of Bologna. Oldoni in *Ahenaeum liguricum*, 380, Perusia, 1680, says about him: "vir morum probitate et scientiarum cognitione nostra aetate celebris, ac Poesiae amator." Oldoni gives (p. 390 sq.) a list of his works, chiefly medical ones, and adds: "multa insuper elaboravit quae nondum impressa in plurimum scrinis asservantur inter quae: *Cosmographiae libri quattuor, Theoricarum coelestium libri tres, Compendium doctrinae sphericae etc.*" It seems that these works remained unpublished; at least they are not listed in the bibliography published by Jean-Baptiste Toselli, *Biographie Niçoise* 2: 287-289, Nice, Paris et Turin, 1860. *Cf.* also Onorato Derossi, *Scrittori piemontesi . . .*, 128 and 216, Torino, 1790; Gio. Bonino, *Biographia medica piemontese* 1: 380, Torino, 1824-1825; Vallauri, *Storia della poesia in Piemonte* 1: 407 and 512, Torino, 1841.

The *Nicetas ortodoxus* is not mentioned by Torrini's biographers, Oldoni and Toselli.

Io. Domenico Cassini. The famous Franco-Italian astronomer and engineer, Jean Dominique Cassini, was born in 1625 in Perinaldo in the county of Nice, became in 1650 professor of astronomy at the University of Bologna and in 1669 was called by Colbert to Paris in order to organize the Royal Observatory which Colbert intended to create, and to become its first director (1672). Thereafter Cassini settled in France, married in 1673, and received, in the same year, the "letters of great naturalization." He died in 1714. Mathematical talent and passion for astronomy were hereditary in the family of the Cassinis. His son, Jacques Cassini, Cassini II (1677-1756), his grandson, Cesar-François Cassini de Thury, Cassini III (1714-1784), his great-grandson, Jacques-Dominique, comte de Cassini, Cassini IV (1747-1845), maintained the tradition which made the name of Cassini synonymous with that of astronomer.

of which the diatribe, after five conclusions, ends nevertheless with this unique conclusion: "that the Sun moves around in the diurnal and annual motion, and the Earth is at rest, I hold firmly, infallibly believe, and openly proclaim, being neither moved by physical nor by mathematical reasons, but by the mere power of faith, the authority of Holy Scriptures and the urge of the Roman See, [by] whose pronouncements, inspired by the Spirit of truth for the teaching of the world, I wish to abide." And after my *Responsio Apologetica*, he resumed his opposition in a *Duplica* which he wrote to me and maintained his previous opinion. Later on, I have more than once had a discussion about this my argument with my friends the most excellent Professors of Science in the University of Bologna, D. Jo. Dominico Cassini and D. Geminiano Montanari widely known by their works published in Italy as well as abroad, but to me very well known and highly esteemed on account of familiar conversations. But then the most excellent D. Jo. Alfons Borelli Professor of Mathematics first in his native town, Messina, and then in the Academy of Pisa, first in a writing and then in his published book the *De Vi Percussionis*, prop. LVIII has attempted to refute my argument against the motion of the Earth. And recently R. Fr. Stefano de Angelis, Professor of Mathematics in the University of Padua, has also asserted, in two dialogues, the inefficacy of either form of my argument; to whom, with my consent, replied D. Michael Manfredi; Father Stefano, once more, inveighed against Manfredi, and also D. Adrian Nasut Gallus in his *Observationibus* p. 49.[349]

[349] I owe to the sagacity and matchless erudition of Dr. Paul-Henri Michel, Conservateur adjoint at the Bibliothèque Mazarine, the identification of the mysterious Adrianus *Nasut* (or *Hasut*, as Stefano degli Angeli calls him, cf. p. 391) with the French astronomer and scientist, Adrien Auzout, who, in 1665, published a *Lettre a Monsieur l'Abbé/Charles/sur le Raggionaglio/ di due nuove osservationi/da Guiseppe Campani/ Avec/des Remarques où il est parlé / des Nouvelles descouvertes /dans Saturne et dans Jupiter*, Paris, 1665.

It is probably this work that Riccioli has in view. He misquotes, indeed, the title, but, as he misnames also the author, it is safe to assume that he has not seen it, but only heard about it from somebody else (as for Stefano degli Angeli he, obviously, only added a misprint to Riccioli's misspelling). In any case, on pp. 17 and 48 sq. Adrien Auzout mentions Riccioli, though he does not criticize his geostatic argument. He, simply, asserts the truth of Copernicanism, its rejection by Giambatista Riccioli and Honoré Fabri notwithstanding.

It is interesting to note that even among his own order the Ricciolian argument did not enjoy universal recognition. Thus the famous Belgian mathematician, André Taquet, rejected it outright, stating that there is not a single valid argument against the motion of the Earth—that of Riccioli being a paralogism— and that we are assured of the Earth's rest only by revelation.

Cf. *R. P. Andreae Tacquet, Antwerpiensii e Societate Jesu, Opera Mathematica*, demonstrata et propugnata a Simone Laurentio Veterani, ex comitibus Montis Calvi, In collegio Societatis Jesu, Lovani, *Astronomiae liber octavus/Varii tractatus Astronomici/tractatus primus, De hypothesi Terrae motae*, 1668. In the introduction to this treatise Tacquet states (p. 321): "Ego quidem minime dubito, quin *Terra in aeternum stet; etenim* firmavit orbem Terrae qui non commouebitur. Unde et libris septem prioribus Astronomiam Universam iuxta Hypothesim Terrae Stantis exposui. Quia tamen haec controversia inter omnes Astronomicas celeberrima est, eam penitus praeterire non est visum. Exponam igitur primo naturam et conditiones huius Hypotheseos, et qui per eam phenomenis satisfiat, deinde quid de eius veritate sentiendum. Utrumque quam potero clarissimē et brevissimē."

Having then (in chap. I and II) given a short but perfectly

Riccioli proceeds then to tell us about his investigations of the fall of bodies in which he claims—and rightly as we know—to have established experimentally the value of the acceleration constant and the validity of the Galilean law of fall (*S* proportionate to t^2).[350] It is on the basis of this law and the fact that the actual speed of descent depends on the (specific) weight of the descending bodies and the resistance of the medium (air) that they have to overcome that he has calculated the actual trajectories of these bodies.[351]

In doing it Riccioli, as we know, had admitted that the gravitational acceleration constant had the same value throughout the world [352] and that the resistance of the medium (air) could be assumed as equally constant throughout the sublunar space, adducing as proof of this experimental "axiom" the fact that bodies fell with the same speed in Bologna as in Florence or Rome.[353]

Quite insufficient as proof, objects Father Stefano; the basis is too narrow.[354] Besides, we don't know anything about gravity and it is quite possible that, just like magnetic attraction, its action weakens with the increase of distance. Moreover, points out the Count, even if its force were constant, the motion downward would not be a uniformly accelerated one during the whole time of the fall because, as Galileo has shown, the resistance of the medium will increase with the speed.[355] Finally, it will even transform the accelerated motion into a uniform one, adds the mathematician.[356]

It is interesting to see that the old problems concerning the mechanism of acceleration continued to be debated in spite of Galileo's refusal to deal with the nature

correct exposition of the system of Copernicus, he continues by asking (chap. III): "*Caput III.* Quid de veritate absoluta motus Terrae sentiendum?" and tells us that R. P. J. B. Riccioli in his "*Almagest.* libro 9 tota sectione quarta quae iustum volumen explet eā tum eruditione tum copia prosecutus est, ut facile omnes hoc in negotio superaverit." Indeed, says Tacquet, R. P. J. B. Riccioli not only quotes all the arguments for the motion of the Earth (49) and against it (77) but also adds a new one.

Chap. IV and V present Riccioli's demonstration and in the Chap. VI (p. 328) its *Paralogismus ostenditur.* The criticism of Tacquet is not original, and not interesting (probably inspired by Mersenne's criticism of Galileo). He states that the trajectory would not be circular but a spiral, and that, if it were circular the fall would last six hours whatever the starting point, which is exceedingly unlikely and moreover incompatible with Riccioli's own teaching and experiences. He concludes therefore that there is no valid argument either for or against the motion of the Earth and that the *Scriptura sancta* is our only reason for rejecting Copernicus.

[350] Cf. *supra*, pp. 351 sq. and my An experiment in measurement, *Proc. Amer. Philos. Soc.* 97 (2): 222–237, 1953.

[351] Cf. *supra*, pp. 352 sq.

[352] As we know this was admitted not only by Galileo but even by Borelli; cf. my La mécanique céleste de J. A. Borelli, *Revue d'Hist. des Sciences*, 1952.

[353] Cf. *Apologia*, 53.

[354] Cf. *Quarte Considerationi*, 51.

[355] *Ibid.*, 52 sq. referring to the Galilean *Dialogue on the Two Greatest World Systems*, p. 75 of the Latin edition.

[356] *Ibid.*

of gravity. Riccioli believes that the acceleration is the result of a new *impetus, impetus productus*, engendered by the gravity of the body at each instant of its fall.[357] Stefano degli Angeli, objecting that a constant force or quality (gravity) cannot constantly produce new effects in an indefinite or infinite number, asserts that gravity, whatever it may be, produces a uniform motion, and he holds that it is not, properly speaking, *accelerated*, but only *de-retarded*—a theory that Galileo, at least the young Galileo, also held.[358] According to this conception, a body thrown up—or simply brought up—to a certain altitude acquires through this elevation a certain *impetus* which is opposed to the impetus of gravity. Thus, in spite of the fact that this latter remains constant and would, as such, produce a uniform motion, it is slowed down and hindered in its action by the acquired, upward driving, one. This latter, however, continuously becoming weakened and used up in resistance, the hindrance that it opposes to the downward motion diminishes, and its speed increases. It is difficult to say whether Father Stefano is really convinced or only uses this theory as a good means of embarrassing Riccioli and making the reader laugh at his expense. Be that as it may, the whole of this first part of Riccioli's *Apology* has no more bearing upon the problem of the validity of Riccioli's main argument against the motion of the Earth than the corresponding part of Angeli's seventh *Dialogue*.

As for this latter, Riccioli tells us once more, that:[359]

If the heavy bodies moved with the diurnal motion of the Earth and, descending, described in the World space the periphery of a circle as stated by Galileo, they would descend uniformly, though maintaining, in the first two or even four seconds of the hour, the proportion of the square numbers—

in respect to the spaces traversed. In this case the acceleration of the motion downward would be apparent only and not real, whereas experience and his experiments have demonstrated that it was physical and real: indeed the impact of the falling body increases with its speed.

Riccioli complains that Stefano degli Angeli has criticized him for assuming that the trajectory of the falling body will be a circle. As a matter of fact, he did not assert it. He only[360]

used this method, in order to make an *argumentum ad hominem* against Galileo who spoke in an abstract way, not descending to any particular case, [which] gave to P. Stefano degli Angeli a remote opportunity to believe that even in the case of the ball let fall from the summit of the Torre degli Asinelli which in the first four seconds of the hour traversed consecutively 15, 60, 135, 240 feet this ball descends by the circular line of Galileo.

Yet it has been a misunderstanding—or worse—on the part of Angeli, as in this case Riccioli did not assert that the ball will *always* follow an exactly circular line. As it matters very little whether this line be circular or not, the difference, in the first four seconds of the fall will be in any case exceedingly small, as well as the difference in the force of percussion arising from the not quite perpendicular direction of the stroke, so small as to be imperceptible. *Ad sensum* therefore the fall would be circular. Moreover, as already stated, in the *Astronomia Reformata* Riccioli has extended his argument so as to include all possible curvilinear trajectories:[361]

Thus it is inopportune to apply a theory which, taken generally and abstractly, is true, and . . . possible, to a case in which it is not verified, as he [Angeli] does on p. 9, that is to the case in which our ball, let fall from the upper stories of the Torre degli Asinelli, traversed in the first second of the hour 15, and at the end of two seconds 60 feet; indeed, we used this experiment in order to prove that the proportion of the spaces of the descent is such as between the squares of the times, and not in order to confirm the circular path.

Father Stefano, of course, maintains his position: the line of descent will neither be a circle, nor any one of the lines alleged by Riccioli, but, as has been demonstrated in the *Secunde Considerationi*, and once more in the *De infinitis spiralis inversis*, a spiral line. Riccioli has probably not read these passages, or, more probably still, has read them, but not understood.[362]

The difference between Riccioli's opinion and that of his critics, pursues Stefano degli Angeli, skillfully using the opportunity to join arms with Borelli, is simply that Father Riccioli does not understand the fact of the relativity of motion:[363]

---

[357] *Cf. Apologia*, 52.

[358] *Cf. Quarte Considerationi*, 54; *cf.* my *Etudes Galiléennes*, Paris, Hermann, 1939, and R. Giacomelli, *Galileo Galilei Giovane*, Pisa, 1949.

[359] *Cf. Apologia*, 40; prop. 9, & 5: "Si grauia cum terrae diurno motu mouerentur e descendende designarent in mundi spatio circuli periferiam à Galileo descriptam: descenderent per illam uniformiter, salua in primis duobus, imo quattuor secundis horariis proportione deita quadratis temporum numeris."

[360] *Ibid.*, 44: "Hac methodo uti, ut ad hominem argumentaremur contra Galileum, qui abstracte locutus erat, non descendendo ad ullum casum particularem, dedimus occasionem remotam P. Stefano de Angelis existimandi, etiam in casu globi ex Asinella Turris uertice demissi, qui primis quattuor secundis horariis

confecit ordinatim pedes 15. 60. 135. 240 globum hunc descendere per circularem lineam Galilei quod minime expressimus."

[361] *Ibid.*: "Et in Astronomia Reformata lineam illam decernimus. Non oportebat ergo doctrinam illam, quae generaliter et abstracte sumpta vera est, et ut mox docebo, possibilis, applicare ad casum quo non verificatur, sicut fecit pag. 9 nempe ad casum, in quo globus noster ex rostris superioribus Asinellae Turris dimissus confecit primo secundo horario pedes 15 et in fine duorum secundorum pedes 60 etc. hoc enim experimento usi sumus ad comprobandum proportionem spatiorum decursorum talem, qualis est inter quadrata temporum, non autem ad comprobandam viam circularem."

[362] *Quartae Considerationi*, 57–58.

[363] *Ibid.*, 68: "MAT. Ma passamo al cap. 5 che principia alla pag. 66 nel quale ricerca se mouendosi la terra, e discendendo il graue, se questi si mouesse vero, e fisicamente per la linea perpendicolare all' ingiù, o pure solo apparentamente. Supponiamo che la terra non giri, como in realtà non gira, e il graue descenda per la perpendicolare all'Orizonte; e lo vediamo

MAT. But let us pass to chap. V which begins at p. 66, in which he investigates whether, if the Earth moved, and the heavy body descended, this would move truly and physically along a perpendicular line downward, or do so only in appearance. Let us suppose that the Earth does not turn, as it does not turn in reality, and that the heavy body descends on the perpendicular to the horizon; we shall see it descend on the perpendicular, and in reality it will descend on it. If the Earth turns and if the heavy body does not descend but remains at rest at the summit of the tower: it will, in reality, move in a circular motion but we shall not see it. Let the Earth turn and the heavy body descend. We shall see it descend on the perpendicular, and in reality it will describe in the World space a curved line, which we shall not see. Now R. P. Riccioli investigates whether in this case the heavy body will descend truly and physically on the perpendicular, as we see it. He names the Sign. Borelli, Cassini and myself who have said that physically and really there are three things: the descent of the heavy body on the perpendicular in virtue of gravity; the gyration due to the diurnal revolution; and from these two motions, the generation of a curve in the world space. P. Riccioli thinks, on the contrary, that this descent on the perpendicular is only apparent, and is not physical and real.

Father Stefano, of course, only sums up what he has said time and again in his *Prime* and *Secunde Considerationi*.[364] Yet he has to repeat his assertion, as Riccioli, indeed, in his *Apologia*, reproduces the same objections that he dictated to Michele Manfredi, objections based upon the old, scholastic conception of motion as a process affecting the body moved. Two motions, he repeats, cannot be combined in the same body,[365] especially when these motions, or the forces that produce them, are essential to the bodies in question. A "mixture" can take place when one of them is essential, and the other not. Thus for instance, when a body is dropped from the top of the mast of a moving ship, its motion down, resulting from the essential intrinsic virtue of gravity, can be combined with the extrinsic motion and virtue received from the moving ship.[366] In the case of the Earth's motion the situation would be quite different: the rotating virtue would be an essential and an intrinsic one.

As a matter of fact, even the mixture of an intrinsic, essential virtue with an extrinsic one is not easy. Thus,

Riccioli reminds us that the trajectory of a cannon ball shot from a gun starts by being rectilinear, and curves down toward the Earth only at the end of its flight, i.e., when the horizontal impetus is spent.[367]

Angeli, of course, denies all that. The rotating motion of the Earth—if the Earth moved—need not be essential. An angel could be entrusted with turning it around.[368] As for the mixture of two or more motions, nothing is easier. They always mix or, if we prefer, nature is perfectly able to hold them apart, so that the effects of the one do not impair or hinder those of the other, as Kepler and Descartes have shown it occurring in the case of the reflection of light.[369]

The cannon ball, therefore, will start falling from the very moment it leaves the gun. Its horizontal motion does not affect or hinder its motion downward. It is the same in the case of the rotating Earth. The motion of the semidiameter upon which the falling body descends toward the Earth (the motion of the Earth) does not affect its motion down. It will, therefore, be physical and real. As for the force of impact, it is only this motion, the motion downward, that has to be taken into account—the horizontal one, common to the Earth and to the body in question, has no effect. You cannot strike a man who runs away from you with the same speed as you are running after him.[370]

The argument of Riccioli is worthless. No more than anybody else has he been able to demonstrate that the Earth is at rest. Indeed it is impossible to do so as in both cases—whether the Earth moved, or not—all the phenomena available to us, all the phenomena observable by us would be exactly the same. To find a difference we should look at the Earth from outside. But we cannot do it.

Stefano Degli Angeli agrees, of course, that the Earth does not move. But it is only faith, faith in the infallible inspiration of the Holy Church that can make us assured of this fact, and not the arguments of Father Riccioli.

Thus we are back, or nearly so, to where we started from. Yet we must not complain. The way that we have trodden was interesting. In studying this long discussion about the trajectory of a falling body *in hypothesi terrae motae* we have been able to gain an insight into the state of physical science and the development of scientific thought at an extremely important moment, the very moment when Hooke and Newton turn their attention to the same problem.

As a matter of fact, this whole story is a *Prelude to Newton.*

---

descender per la perpendicolare, e in realtà per essa descende. Giri la terra e il graue non discenda, ma sia immobile nella sommità della torre; egli si mouerà realmente di moto circolare ma non lo vederemo. Giri la terra, e il graue descenda; noi lo vederemo descender per la perpendicolare, e in realtà nel spatio mondano descriuera una linea curua, che noi non vederemo. Ricerca hora il P. Riccioli se in questo caso vera, e fisicamente il graue descende per la perpendicolare, come lo vedemo. Nomina li Signori Borelli, Cassini, e me che habbiamo deto fisica, e realmente darsi tre cose, descender il graue per la perpendicolare in virtu della grauità; girare per la riuolution diurna; e da questi due moti generarsi la curua nel spatio mondano. Al contrario pensa il P. Riccioli, che questa scesa per la perpendicolare sia solo apparente, e non fisica e reale."

[364] Cf. *supra*, pp. 361, 364.
[365] Cf. *Apologia*, 68 sq.
[366] *Ibid.*, 70.

[367] *Ibid.*, 77, 88; cf. *supra*, p. 370.
[368] *Quartae Considerationi*, 20.
[369] *Ibid.*, 72, 73.
[370] *Ibid.*, 77, 81. Stefano degli Angeli reproduces an example of Borelli. It is curious that Father Stefano does not recognize that in his opposition against the attribution to the Earth of an "extruding faculty," he, as a matter of fact, takes side with Riccioli against Borelli.